AFRICAN NATURALIST

AFRICAN NATURALIST

The Life and Times of
Rodney Carrington Wood, 1889–1962

David Happold

David Happold

Book Guild Publishing
Sussex, England

First published in Great Britain in 2011 by
The Book Guild Ltd
Pavilion View
19 New Road
Brighton, BN1 1UF

Typesetting in Garamond by
Ellipsis Books Limited, Glasgow

Printed in Great Britain by
CPI Antony Rowe

A catalogue record for this book is available from
The British Library.

ISBN 978 1 84624 555 8

To my wife Meredith, who has shared my enjoyment of Africa and assisted me in so many ways, and to our two children – Karen Lena and Jonathan – who were born in Africa.

CONTENTS

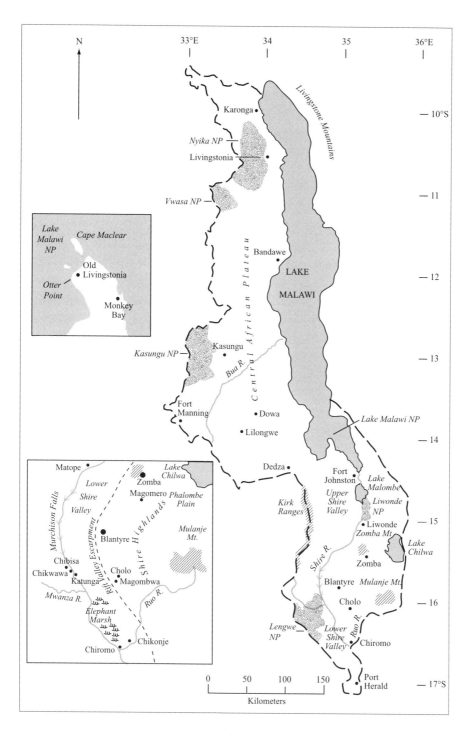

N

33°E   34   35   36°E

Karonga

— 10°S

Nyika NP

Livingstonia

Vwasa NP

— 11

Bandawe

LAKE
MALAWI

— 12

*Lake
Malawi
NP*

Cape Maclear

Old
Livingstonia

*Otter
Point*

Monkey
Bay

Kasungu NP   Kasungu

— 13

Bua R.

Fort
Manning

Dowa

Lilongwe

Central African Plateau

Livingstone Mountains

Lake Malawi NP

— 14

Dedza

Fort
Johnston

*Lake
Malombe*

*Upper
Shire
Valley*

*Liwonde
NP*

Liwonde

Zomba Mt.

Zomba

Blantyre *Mulanje Mt.*

Cholo

*Kirk
Ranges*

Shire R.

— 15

— 16

*Lake
Chilwa*

*Lengwe
NP*

*Lower
Shire
Valley*

Chiromo

Port
Herald

— 17°S

Matope

*Lower
Shire
Valley*

Zomba

Magomero

*Lake
Chilwa*

*Phalombe
Plain*

Shire Highlands

*Mulanje
Mt.*

Murchison Falls

Rift Valley Escarpment

Blantyre

Chibisa

Chikwawa

Katunga

Cholo

Magombwa

Mwanza R.

Ruo R.

*Elephant
Marsh*

Chikonje

Chiromo

0   50   100   150

Kilometers

ACKNOWLEDGEMENTS

Many people and organisations have helped during my research for *African Naturalist*. I am most grateful to many correspondents (listed here alphabetically) who sent me books, letters, reminiscences, and advice: Don Baring-Gould, Chris Barrow, Lyn Barrow, Lady Beit, Debbie Bellairs, Harry Boles, Don Broadley, Peter Charlton, Desmond Clark, Don Currie, Athalie Ducrotoy, Alan Foot, Viv Fynn, David Hancock, Fred Handman, Helen Harrison, Tony Hooper, Peter Jackson, Alan Kohn, Bill Lamborn, Ian Lane, Don Mackenzie, Brian Marsh, Anne Metcalf, Aubrey Michel, Brian Morris, A. Nageon de Lestang, Nadege Nageon, John Ness, Pat O'Riordan, Don Potter, R. Purdy, Arthur Schwarz, Gwynneth Scrivener, David Stuart-Mogg, Peter Taylor, Des Tennett, John Trataris, Gwyneth Walker, George Welsh, John Wilson, Vaughan Winter, and R. Wright.

I examined many of Rodney Wood's collections, and I am grateful to the curators and their staff who allowed me to examine Wood's specimens or sent me information on Wood's collections: Paula Jenkins (mammals – Natural History Museum, London), Robert Pryce-Jones (birds - Natural History Museum, Tring), Alwyne Wheeler (fish - Natural History Museum, London), Woody Cotterill and Karen Donnan (butterflies - Natural History Museum of Zimbabwe), Dick Kilburn (shells - Natal Museum, South Africa), Kenneth Parkes (birds - Carnegie Museum, Pittsburgh, USA), Timothy Matson (Cleveland Museum of Natural History), and John Rawlins (invertebrate zoology - Carnegie Museum, Pittsburgh, USA). John Thackray (Archivist, Natural History Museum, London) sent me copies of Wood's extensive correspondence with Oldfield Thomas and George Boulenger.

During the research for this book, I visited many places where Rodney Wood lived. At Ardvreck School, Michael Kidd (Deputy Headmaster) showed me around the school and gave me details of the school when Wood was a student there; Rita Boswell (Archivist) provided me with Wood's reports for the years when he was at Harrow; and Ronald Brooks (Archivist) conducted me around Michaelhouse School and provided me with reports and letters about the school in the 1930s. Information on Wood's years with the Boy Scouts in England and Canada was provided by the Scouts Association (UK) and by Harry Boles in Canada. In the Seychelles, Julian Durup (Archivist, National Archives) found all the documents relating to Wood's purchases of land and gave valuable information on the Seychelles in the 1920s and 1930s. In Malawi, at the National Archives in Zomba, I was able to see Wood's *curriculum vitae,* which he wrote when applying for the post of Chief Game Warden; and - thanks to Michael and Tete Roberts - I was able to stay in Rodney Wood's house at Magombwa and walk through the forests and tea plantations where he and the Westrops had lived and worked for so many years.

During the course of this study, I interviewed several people who knew Rodney Wood and/or who had lived in Nyasaland in the 1940s and 1950s. Their reminiscences provided a wealth of information about Wood himself and about life in Nyasaland: Richard Westrop (son of Arthur Westrop), Dick Mullon and Des Tennett (both of whom lived in Nyasaland in the early days), Bill Lamborn (son of W. A. Lamborn, Chief Entomologist during Wood's early days), Viv Fynn (a distant relation of Wood's on his mother's side), and Valerie Westrop (wife of Richard Westrop). In the Seychelles, I met and interviewed Marcel Mathiot (who, as a young boy, spent a lot of time with Rodney Wood on Cerf Island), Lena Nageon de Lestang (one of the many members of the Nageon family who were great friends of Wood's), and Henri McGaw and Doris Johnston (both of whom provided details of life in the Seychelles at Wood's time). Alan Foot organised a 'Wood Day' in Harare where I met many people who knew Wood and were his contemporaries in Nyasaland.

I am grateful to the Syndics of the Cambridge University Library for permission to quote from the Rodney Wood Diaries (Ref: GBR/0115/RCMS 128), to the African Lakes Corporation (through Don Mackenzie) to reprint old photographs, and to The Wildlife Society of Malawi for access to Wood's field notes on his bird collections.

Valerie Westrop kindly gave permission to reproduce the delightful caricature of Rodney Wood.

This study would not be so comprehensive without the special contributions of six people, to whom I owe immense gratitude. Terry Barringer of the Cambridge University Library alerted me to the Rodney Wood Diaries and allowed me to quote extensively from the diaries; John Weldon gave me a copy of Arthur Westrop's book *Green Gold*, a goldmine of information about Wood; Vera Garland - formally Librarian of the Society of Malawi in Blantyre - provided me with many important papers and documents; Marcel Calais, Wood's great friend and confidant on Cerf Island in the Seychelles, shared his memories with me; Rosemary Lowe-McConnell shared her memories, diaries and stories about the wonderful times when she lived at Monkey Bay and when Wood helped her with her research; and my wife, Meredith, was my companion during our many travels while gathering material for this book.

PROLOGUE

It had been a scorchingly hot dry day, but at long last the sun was setting over the Kirk Ranges in southern Malaŵi. It was April 1994, and my wife and I were nearing the end of a wonderful year studying the small mammals of this beautiful African country. We were on our way to Chiromo, a place that had been in our imagination and in our thoughts for nearly ten years. For much of our time in Malaŵi we had been living on a farm near Zomba in the highlands, but our work had also involved many days of fieldwork to places which were of particular interest to our studies. One of these places was Chiromo, on the banks of the Shire River, where a collection of very interesting and rare species of bats had been assembled during the First World War by an Englishman named Rodney Wood. In the eighty years since then, many environmental changes had taken place in this part of Malaŵi and we were anxious to find out whether Wood's special bats were still living in Chiromo.

We drove slowly along the bumpy laterite road, passing little groups of Africans chattering loudly as they walked home from work. Others were cycling sedately along the railway line beside the road – wind-blown soil had filled up the spaces between the sleepers, so it was much easier to ride a bicycle here than on the road. We crossed the metal bridge, with the wheels straddling the railway lines; below, the Shire River flowed evenly and slowly towards the Zambesi river and eventually the Indian Ocean. We had been told that there wasn't very much in Chiromo now. We drove along a dusty avenue of old trees. Among the unkempt savanna grasses, we could see the remains of once-stately homes, with sections of walls missing, broken bricks on the ground, and the cracked concrete floors. Within five minutes, we realised we had travelled right through Chiromo, and that there was virtually nothing left of this famous township. It needed only a few years of African heat, rain and flood to eliminate most of the traces of human occupation.

We retraced our steps and found a gateway, not far from the bridge, that we hoped would lead to the Ruo River which flows into the Shire at Chiromo. We walked in, and much to our surprise, found a large two-storey house which, in the old days, had been occupied by the manager of the British Cotton Growing Association. It was set amid tall, mature trees, with the Ruo river flowing between steep banks at the back. The house had a tin roof, verandas and shutters; it would once have been a beautiful home with a lovely garden, but now it was neglected and unoccupied. On the bare soil nearby, the watchman had lit a small fire and was preparing his supper of maize meal and fish. Another man was carrying two freshly caught lungfish from the Shire River. A little group of children came across to see us – *wazungu* (white people) were a rare sight in Chiromo now. We explained why we had come, and asked their permission to set our special nets for catching bats near to the river. By the time it was almost dark, we had set three nets, placed head torches on our foreheads, and checked that the bags, gloves, notebooks, echo-location equipment and all the other paraphernalia required for our work were to hand. Thankfully there was no wind, and we knew it would be a clear starlit night with no moon – perfect for catching bats! The air was still and it was very quiet, with just a few murmurs of conversation from the watchman and his friends around the fire, the occasional bark of a dog in the distance, and the quiet hoots of Scops Owls.

There is always an air of excitement and anticipation while catching bats at night. One never knows what may fly into the nets; there may be no bats at all, or so many that one is hard pushed to remove each one quickly, or there may be something very special . . . On this particular night, we were more excited than usual because if we found any of Wood's special bats, we would be the first biologists to see them in Malaŵi for over eighty years, and (most importantly) we would know that they had survived despite the huge environmental changes in the Lower Shire Valley. By about 6.30 p.m. it was almost dark, and we could see the silhouettes of bats flying near the house and around the trunks of the big trees. Darkness comes very quickly in the tropics, and within a few minutes, it was pitch black. We stood quietly by the nets; every minute or so, we turned on our head torches and flashed the light along the length of the net to see if any bats had been caught. Sometimes we could hear a slight swishing sound when a bat flew into the net. Any bat that was caught was quickly and carefully removed, examined briefly, and placing in a cotton collecting bag. It turned out to be a wonderful night: in about two hours we caught twenty-five bats belonging to nine species. Some, as we expected, were common species which we had caught elsewhere in Malaŵi, but three species were Wood's special bats. We could hardly believe our good luck! First there was a species called *Scotoecus albofuscus*; most of its body is brown, but the wings are white and translucent, quite unlike any other bats we had encountered. Then there was *Glauconycteris variegata*, a beautiful browny-orange bat with orange wings which are finely etched with thin brown lines, reminiscent of the delicate venation on a leaf. Finally, there was *Eptesicus rendalli*, a brownish bat with a white underside and white wings. We took the nets down at 8.30 p.m., but because there was so much to record about these bats, we did not get to bed until just after 1 a.m the next morning. For us, it was a truly magical evening! We stayed for two more nights and caught many more bats, but the first night was undoubtedly the most exciting of all.

One of the essential requirements of fieldwork on mammals in Africa, or indeed anywhere in the world, is a thorough knowledge of what is already known about the species. This involves many days (and weeks) of work reading scientific papers, writing to colleagues, and planning a reasonable and achievable research project. During our reading, we came across one of the earliest papers on the mammals of Malaŵi. It was published in 1922 in the long-established and respected English journal *The Annals and Magazine of Natural History*, and its title was 'On a Collection of Mammals from Chiromo and Cholo, Ruo, Nyasaland, made by Mr Rodney C. Wood, with Field-notes by the Collector'. The author, P. S. Kershaw, of the British Museum of Natural History in London, had received the collection, identified the specimens, and written the results for publication. But the most remarkable feature of the publication was the extensive notes about the habits of each species that Rodney Wood had sent with the specimens. This was most unusual; many specimens collected in those early days were woefully lacking in information: cryptic notes such as 'Africa' or 'Kenya 1907' were often deemed to be adequate! Kershaw had had the good sense to include most of Wood's field notes – in fact much of the paper comprises the field notes rather than Kershaw's identification. We realised from this, and from the beautifully prepared specimens made by Rodney Wood which are still held in the museum, that he must have been a very remarkable and observant person. Some of the species of small mammals obtained by Wood, especially the bats, have rarely been caught in Africa and are known, within Malaŵi, only from Chiromo.

After reading about Wood's collection, we decided that it would be well worthwhile to visit Chiromo to ascertain whether any of these rare bats were still living there. During the 80 years since Wood obtained his specimens, there have been many environmental changes in the Lower Shire Valley, primarily because the human population has increased tenfold. Large areas of savanna woodland have been replaced by farms and by extensive cotton and sugar cane plantations. Trees have been cut for building and firewood. Many of the habitats which are required for mammals, and for many other species, have either shrunk in area or been eliminated. Hence, we surmised, it would be of interest to find out whether the small mammals have been adversely affected or, indeed, have disappeared altogether. But it was not until April 1994 that we were able to get to Chiromo, and spend those three exciting days and nights looking for 'Wood's species'.

Kershaw's publication did not tell us anything about Rodney Wood himself just that 'this interesting collection of beautifully prepared specimens is the result of the labours of some years'. Long-term residents in Malaŵi were not able to enlighten us very much either; this was hardly surprising since Wood made his collection during the First World War. However, we did obtain some tantalising snippets of information, such as: 'I think he was Chief Forester at one time'; 'He was a church minister'; 'Wasn't he a tea planter?'; 'He was known as "butterfly-brain"'; 'Didn't he own an island in the Seychelles?'; and 'I think he was Arthur Westrop's friend.' This motley selection of improbable comments was the beginning of what proved to be a long and fascinating quest. During the three months between visiting Chiromo and our departure from Malaŵi, we made three significant finds. First, if Wood really had been Chief Forester, we thought there should be some files about him in the

National Archives in Zomba. After a day of searching, we unearthed a single file begun in 1928 when Wood applied for the post of Game Warden in Nyasaland. It included his application for the post, which gave his date of birth, the schools he attended, his experience and qualifications – all wonderful fodder for future investigation. Second, we were told that the *Nyasaland Journal* had published an obituary for Wood in 1962. The obituary, written by Arthur Westrop, contained even more information and further points for investigation. Third, we heard about a remarkable book called *Green Gold*, which was a privately printed autobiography by Arthur Westrop. The book is virtually unobtainable now, but when we were lent a copy we found that Wood was mentioned on many occasions, and we realised that Wood and Westrop were friends for the best part of forty years. To begin with, we were only interested in finding out more about Wood's collections – what he collected, when he collected, where the collections are housed now, and the value of his collections to science. But in time, we became more interested in the places where Wood lived and what living conditions were like in those far-off days. And finally, we became interested in Wood as a person: What was he like? What did he do? And what did his friends think of him?

As the study enlarged in scope, it was necessary to read many books which recorded the exploits of famous travellers in this region of Africa. The most famous, of course, was David Livingstone. His expeditions on the Zambesi and Shire rivers and northwards to Lake Nyasa from 1858 to 1863 are best known for their efforts to curtail the slave trade, to bring Christianity to central Africa, and to discover the source of the major rivers and lakes. Accompanying Livingstone were many remarkable men who, until recent times, have not been given sufficient credit for their contribution to the 'opening of Africa'. John Kirk and James Stewart, for example, kept fascinating diaries during their travels that included many references to the natural history of the region. Kirk and Stewart were followed by others who also recorded their observations of plants and animals. It is a lasting tribute to their powers of observation and energy that such people were able to write notes and collect specimens while enduring unbelievable hardships due to disease, climate and isolation. These books provided a background to Nyasaland before Wood first arrived in the country in 1914.

Finding out about Wood himself has been a fascinating experience involving visits to many places in England, Scotland, Malaŵi (as Nyasaland is now called), Zimbabwe, South Africa and the Seychelles, as well as visits to museums, and correspondence with many people who knew Nyasaland in colonial days and some who knew Wood personally. He lived during one of the most memorable periods of African history. He was born in 1889, the same year that Harry Johnston travelled up the Shire River for the first time and two years before Johnston became the first Commissioner to British Central Africa. He died in 1962, two years before Nyasaland became the independent nation of Malaŵi under the leadership of Dr Hastings Banda. Thus, his life spanned virtually all of the colonial period, and he was witness to one of the most remarkable periods of change on the African continent.

Searching for details of Rodney Wood's life involved a lot of correspondence – a veritable treasure hunt. And like all treasure hunts, there were times of great elation when, for example, a fascinating letter or photograph arrived, and times of frustration when a line of investigation petered out to a dead end or when some potentially useful person I had written to never bothered to acknowledge my letter. Often, there was a chain of events whereby one person replied, suggesting that I write to someone else; sometimes this continued through three of four correspondents until, at the end, I reached a correspondent who was a wonderful goldmine of information. There were also some quite extraordinary pieces of luck. On one occasion, I received a letter from a correspondent in Zimbabwe with a rather cryptic postscript: 'I think there is person called Ireland who works for Qantas whose mother may be able to tell you something about Rodney Wood.' It was not much to go on, but worth investigating – so I telephoned the local office of Qantas (the national airline of Australia) and enquired whether there was anyone with the surname of Ireland who worked for them. After an hour or so, and of talking to people in offices all over the country, I was given the names of six Irelands. I wrote to all of them, and soon received a letter from Captain Ireland saying it was his mother I was looking for, and that she had known Rodney Wood and was expecting to hear from me. The outcome was a wonderfully long letter that arrived from New Zealand, full of reminiscences of Wood during the time he lived in Rhodesia. On another occasion I wrote to Michaelhouse School in

Natal, South Africa, because Arthur Westrop, in his obituary of Wood, mentioned briefly that Wood had taught natural history and French at the school. I enquired whether there were any records about Wood's activities while he was at Michaelhouse. The reply from the school archivist, besides including photocopies from the school magazine for the period, mentioned that he had received a letter very similar to mine from Tony Hooper in Cape Town three months previously. So began another set of correspondence. It turned out that Tony Hooper was distantly related to Rodney Wood (through the younger sister of Wood's mother); moreover, this led to finding another relation, Vivienne Fynn, who lives in Zimbabwe, and who is descended from the elder sister of Wood's mother. Many other stories could be told, all equally fascinating and improbable.

This is a different story to most written about Africa. Many books on Nyasaland and Malaŵi deal with history, politics, administration, geography, missionary work, ethnography and hunting. There are very few which describe the day-to-day life of Europeans who went out to settle or work there. The best documented accounts of life in Nyasaland in the early days of the colony are those written by missionaries. But most of these accounts were written prior to the First World War, and there are very few for the years when Rodney Wood lived in Nyasaland. In many ways, life in Nyasaland in the early years of the twentieth century was difficult: disease, especially malaria, was widespread; shops and facilities were few and far between; and most people, except those in the towns, were isolated and had to rely on their own resourcefulness. But the rewards outweighed the difficulties: there was freedom, a sense of adventure, satisfaction in doing something worthwhile, the forging of deep friendships, and the pleasure of being able to experience the fascination of Africa. Rodney Wood was not typical of the average Englishman in Nyasaland; he was far too interested in the environment around him, and he spent much of his time collecting and identifying the animals he encountered around his homes and during his travels, many of them hardly known to the scientific world at that time. Many of the snippets of information which we were told at the beginning of our quest proved to be untrue, or only partly true. He was not a church minister, nor a Chief Forester, but he was a tea planter, did own a piece of land in the Seychelles, and was passionately fond of butterflies. Also, he has emerged as one of the most knowledgeable of naturalists to have lived in this part of Africa. Anyone who knew him well was impressed by the breadth of his knowledge; many stories are told of a walk with him through the African bush, accompanied by a fascinating commentary about butterflies, birds, vegetation, Stone Age artefacts, and anything else that happened to be encountered. Although some people found him to be rather offhand and remote, to those who knew him well, he was an inspiration, a brilliant naturalist, an educated and charming person, and a staunch friend. He also had the ability to be at home with people from all stations in life: lords and ladies, governors, admirals, farmers, fishermen, scientists, and little children.

This, then, is the story of Rodney Wood: his life in Africa for nearly fifty years, his collections, his travels, and his contributions to the zoology of Africa.

Chapter 1

CHILDHOOD IN BRITAIN: 1889–1909

Inverness Terrace, London N2, is a wide, straight street which runs northwards from the Bayswater Road. It is lined on both sides by tall, elegant terraced houses built during the Victorian era. The houses face directly onto the pavement and a short flight of steps leads up to the front door. Each house has five floors: a semi-basement protected from the pavement by a low metal paling fence, three main floors, and a fourth attic floor under the roof tiles. Particularly noticeable are the large windows which allow plenty of light into the rooms and which vary in design on each floor. The windows on the first floor are tall, rather like French windows, and those above the porch open onto a small balcony. The windows on the second floor are rectangular with a semicircular pane at the top, and those of the top two floors are also large but conventional. The pavements are wide, with large plane trees spaced regularly along each side of the street. Although by the 1990s most of the houses had been converted into hotels and offices, they must have been very gracious family dwellings in the 1800s and early years of the 1900s. It was here, at 124 Inverness Terrace, that Rodney Carrington Wood was born on 10 October 1889.[1]

Rodney Wood's father, Alexander John Wood, was born in Newcastle-upon-Tyne in January 1860.[2] In time, he became a wine merchant, following the family tradition. Wood's mother, Edith Mary Carrington, was born in Australia, and was one of three daughters born to Charles and Jessie Carrington. Charles had emigrated from Lincoln to Australia, as had Jessie, who was born in London. They married in Brisbane in 1859 and produced three daughters, all of whom were born in Brisbane: Jessie Frances in 1860, Edith Mary in 1861, and Ida May in 1866.[3] Charles must have been a man of some eminence, because he was described on Edith Mary's birth certificate as 'Clerk to the Prime Secretary' in Brisbane, and later in life, on his daughters' marriage certificates, his 'rank or profession' was given as 'Gentleman'. The Carrington girls must have been a stunning trio. Still in existence is a painting on porcelain of Jessie Frances at about the time she was married: she was a striking girl with a rounded face, brown hair, blue eyes and a kind, wistful look. Edith Mary is said to have been extremely beautiful.[4]

At some time between 1879 (when Jessie was married in Sydney at the tender age of 19) and 1887, the Carrington family left Australia and sailed back to Britain. They settled at 87 Inverness Terrace, from where Edith Mary was married to Alexander John Wood on 15 February 1887.[5] Edith Mary's mother and her two sisters, as well as Alexander John Wood's father, were at the wedding in Christ's Church, Paddington. The newly married couple moved into 124 Inverness Terrace, just down the street from Edith Mary's family, and Rodney was born 20 months later.[6] The Wood family was probably quite typical of middle-class families at this time: the national census of 1891 records that six people were living at Number 124: Alexander John, Edith Mary, Rodney, a cook, a nurse-girl and a housemaid.[7]

The marriage was not a happy one. Before Rodney Wood had reached the age of six, his mother had left his father. Alexander Wood then married Emily Blanche Wilde, on 2 February 1895.[8] In the late Victorian era, when the standards of socially acceptable behaviour were much stricter than they are today, separation, divorce and having illegitimate children were frowned upon and, if possible, not recorded on documents. The marriage certificate records Alexander Wood's 'condition' as a widower and his profession is given as 'merchant'. This was not strictly true because Edith Mary was still alive; she eventually married another man and returned to Australia. Emily Blanche was recorded as a spinster (although she already had a daughter born out of wedlock) and as being without any profession although, in fact, she was an actress. She, like Edith Mary, was extremely beautiful, but the fact that she had an illegitimate daughter caused a great scandal within her family.[9] Emily's daughter, Mary Deidre Creagh Wilde, came to live with the Wood family; initially she took her surname from her father and from her mother, although later she tended to use her stepfather's name, calling herself Mary Deidre Wood. Although she was a stepsister to Rodney Wood, there was no close relationship between them; however, by one of those strange quirks of history, they did meet again in South Africa many years later. Alexander Wood's second marriage did not last long either; Emily left him, and for the remainder of his life he lived by himself. There is a great sense of pathos in this family saga: Alexander married two very beautiful women, but both left him after a few years. Although Rodney was brought up in good homes, and went to first-class schools, he would have received very little motherly love and attention. How this may have affected him is unrecorded; however, he never married and never talked about his family; instead, he turned his enormous reserves of energy to other pursuits.

Alexander Wood moved from Inverness Terrace to 36 Abercorn Place in St John's Wood probably during the years when Rodney was at school.[10] St John's Wood is a wealthy upper-class district of London. The houses are large and gracious, many of them surrounded by good-sized gardens. By moving from Inverness Terrace, Alexander Wood was moving up the social scale. Like many people in those days, he did not own his house; instead, he rented it from Harrow School, which at that time owned many properties in London. Number 36 was a detached house with three floors, including the attic floor, and had a garden at the front and back. All the houses of Abercorn Place were built in a similar way, suggestive of a pleasant, rather well-to-do lifestyle. Plane trees and chestnut trees line either side of the street and provide a green and pleasant outlook from the houses. Close by, also in Abercorn Place, is the stone parish church of St Mark's, and on the corner with Lanark Street is the local pub, The Lord Elgin, built on the site of an old coaching inn of long ago. After the bustle and noise of central London, St John's Wood was, and still remains, an oasis of quiet where birds and butterflies frequent the flower-filled gardens.

Alexander Wood was a director of 'Alex. Wood and Campbell & Co', wine merchants and wine shippers, of 62 Crutched Friars in the east central part of London.[11] Exactly when Alexander Wood founded his wine business is uncertain, but it was listed in the London Trades Directory for 1901 and continued as a going concern until 1936. Crutched Friars (the name derives from crossed or crouched friars, and was the site of a monastery dissolved by King Henry VIII during the Reformation) is a maze of little streets just north of the river Thames, upstream from Tower Bridge. Contemporary photographs of the street suggest it was dull and sombre, full of commercial businesses, a monotonous line of brick buildings, three or four storeys tall, each with a short flight of steps to the front door. The windows looked like those of ordinary houses – there were no large glass shop fronts, although a few windows had the name of the business etched on the glass. At one point, there were double gates leading into the 'Port of London Authority: Crutched Friars Warehouse,' where import and export goods were stored. Alex. Wood and Campbell shared Number 62 with the firm of Benscher Martin, forwarding agents. Other businesses in Crutched Friars, just after the turn of the twentieth century, included a motley collection of cork importers, wine merchants, tea blenders, tobacco brokers, engineers, furnishers, and the Ashton Valve Company. Wine importation must have relied on good communications; as early as 1906, Alex. Wood and Campbell are listed with a telegraphic address, 'Skeelam', and a telephone number, 'Avenue 2460' (in those days, telephone exchanges were given names, and connections were made through an operator).[12] Alexander Wood must have made a success of his business, and hence he was able to afford to live in a large house in a preferred part of London, and had sufficient funds to send his son to expensive schools.

Rodney Wood's childhood days were during a time of great social change. At the end of the nineteenth century, transport around London was primarily by horse-drawn carriages. These varied in size from small two-wheeled hansom cabs drawn by a single horse to large four-wheeled double-decker carriages drawn by two large horses with an outside semicircular staircase at the back. The wheels of the carriages were made of wood, with a thick iron rim around the outside that protected the wheel from excessive wear. One often forgotten aspect of horse-drawn vehicles is the large volume of horse droppings deposited on the roads, and the clatter made by so many wheels on the uneven surfaces. Special cleaners with trays and brushes attempted to keep the roads clean, but the smell and flies, especially in summer, were an everyday feature of London life. London streets bustled with carriages and cabs, horse-drawn carts full of produce, and pedestrians who often walked on the roads rather than the pavements. By the time Wood was in his late teens, the first motorised buses were replacing the horse-drawn carriages, and motorised taxis were replacing hansom cabs. Advertising on public transport was as widespread then as it is now, 100 years later; many of the larger carriages were emblazoned with advertisements for Nestles Milk, Pears Soap, Bovril, Colman's Mustard, or Maples Furniture Store.[13]

Dress in late Victorian and Edwardian days was very formal. Photographs of typical street scenes show dark-suited men wearing bowler hats or top hats, and working-class men with waistcoats and flat caps. Ladies wore long skirts reaching to the ankles, jackets and blouses reaching to the neck and wrists, and either a bonnet or a wide-brimmed hat. Children were suitably covered from top to toe. During Rodney Wood's childhood days, the streets and pavements would have been full of tradesmen and hawkers selling and delivering their products: milkmen, flowermen, bakers, butchers, knife-grinders, muffin men, rabbit-sellers, gingercake men, and ice-cream men. Musicians, organ-grinders (sometimes with performing monkeys), speakers, and Punch and Judy shows provided much-needed entertainment.

The great London parks – Hyde Park, Kensington Gardens and Regent's Park – were well established when Rodney Wood was a small boy. He would only have had to walk a short distance southwards along Inverness Terrace to reach Kensington Gardens and Hyde Park. Here the wide grasslands, well-developed trees, lakes and flower gardens would have been a real delight to a budding naturalist. The parks, then as now, were thick with lime, oak, plane and chestnut trees, and the lakes and ponds were surrounded by willows and rushes. Many species of birds are recorded from the London parks; Rodney Wood's lifelong interest in birds may have begun as he watched the mute swans, ducks, gulls, moorhens, coots, wagtails and pigeons there. Nearby to his second home, in Abercorn Place, is Regent's Park with its canal and zoological garden. The zoo has always housed many species of African mammals and birds, and visits to the zoo were a source of interest and delight to Victorian and Edwardian Londoners. Many of the species on display were relatively new to science, and anything from Africa was regarded with great curiosity.

Rodney Wood did not spend all of his childhood in London. His formal education was at boarding schools, so for much of each year between the ages of nine and seventeen he lived away from home. In

May 1899, he travelled north to Scotland, to Ardvreck School at Crieff, a small town sixteen miles west of Perth.[14] Why his father sent him to this school is uncertain; Crieff is a long way from London and there does not seem to be any family connection with this part of Scotland. Undoubtedly, he travelled from London to Crieff by train. By the 1890s, train travel had been in existence for the best part of sixty years and was the established method of long-distance travel. Motor cars were rarely used except within towns and cities. In London, the large railway stations of St Pancras, Euston, King's Cross and Waterloo were huge buildings, built and furnished in the opulent style of the late Victorian period, complete with hotels and gracious restaurants. The railways were run by several independent companies, each of which owned their own engines and rolling stock. To travel between London and Crieff, Wood had the choice of the 'West Coast' route (through Birmingham, Manchester, Carlisle and Glasgow) or the 'East Coast' route (through Nottingham, York, Newcastle and Edinburgh). The 'West Coast' route was operated by the London and North Western Railway (LNWR) and was renowned for its comfort and punctuality, while the 'East Coast' was operated jointly by Great Northern Railway, the North Eastern Railway and the North British Railway, and was well known for the speed of its trains. In the 1890s, there was intense rivalry between these companies, each vying for the fastest, most comfortable service. In 1895, the rivalry was so intense that the two companies staged a race between London and Aberdeen: on 21 August, an East Coast train travelled the 523 miles from London to Aberdeen in 520 minutes; and the following night, a West Coast train travelled the 540 miles in 512 minutes.[15] The journey to Aberdeen now, one hundred years later, takes 430 minutes!

The journey from London to Crieff must have been a great adventure for young Wood. Departing from King's Cross or Euston station, he would have been part of the noise and bustle of passengers finding their places on the train, and the many porters wheeling their trolleyloads of trunks and suitcases. If he went up to the far end of the station (as most small boys loved to do), he would have looked up at the engine driver and the fireman stoking the boiler with coal from the tender, and felt the intense heat wafting from the boiler door. No doubt he would have noticed the name of the engine – perhaps

The Flying Scotsman – and marvelled at its huge size and the vast amounts of smoke and steam belching noisily from the chimney and steam valves. The carriages were well appointed – primarily because travellers demanded comfort and service and because there was such competition between the rival railway companies. Each compartment seated six or eight passengers; the bench-like comfortable seats faced each other and colour pictures of the countryside along the route lined the walls above the seats. A door led into a corridor along one side of each carriage so that passengers could pass from one carriage to another.[16]

Train travellers in the last years of the nineteenth century and early years of the twentieth century could plan their journey by consulting *Bradshaw's General Railway and Steam Navigation Guide for Great Britain and Ireland*, which was published several times each year. It was a thick book, each page filled with columns of timetables and printed in such small print that a magnifying glass was needed to read it. There were many trains each day from London to Scotland, so Wood had a large choice. If he travelled by the East Coast route, the best train allowed him to reach Crieff in one day. The train left King's Cross at 5.15 a.m. and arrived in Edinburgh at 3.40 p.m.; another local train, of the North British and Caledonian line, left at 4.05 p.m. and arrived at Crieff at 6.30 p.m. An alternative was the overnight train which left King's Cross at 10.30 p.m. and arrived in Edinburgh at 8.58 the following morning; the local train departed at 11.17 a.m. and arrived in Crieff at 3.10 p.m. A single ticket in 1900 from King's Cross to Edinburgh cost £2.17s.6d (1st Class) or £1.12s.6d (3rd Class), and a return ticket cost £5.9s.6d (1st Class) or £3.2s.8d (3rd Class).[17]

Although the journey was undoubtedly a long one for a small boy, looking out of the window would have made the journey of great interest to a budding naturalist. In the 1890s, the scenery would have been rather different to that seen by today's traveller. There were extensive woodlands, lots of hedgerows, many cows and sheep grazing in unimproved pastures, numerous small fields, horse-drawn ploughs and harvesters, stooks of cut wheat and barley in the summer, canals with barges, and only a few roads with horse-drawn carts and carriages; there were no cars, no motorways, no large electricity pylons, few bridges, and no plastic or tarmac. Travellers expected to eat well during the long train journey. Maybe young Wood took sandwiches, but he could also have walked down the corridor to the dining room, which would have been beautifully laid with tablecloths, glass and silver. A poster, published in 1900 by the Great Northern, North Eastern and North British Railways, gives some idea of what was available:

Luncheon, dinner and other refreshments will be served en route, the charges being as follows:

LUNCHEON
served in the down train between Kings Cross and York, and up train between Edinburgh and Darlington. First Class 2s 6d Third Class 2s 0d

DINNERS
First Class 3s 6d Third Class 2s 6d

TEAS
served between King's Cross and Edinburgh at any time on the journey. Tea or coffee, with roll and butter 6d. Other refreshments at buffet charges as per day of fare.

LUNCHEON BASKETS
containing hot or cold beverages are provided at the refreshment rooms at King's Cross, Peterboro', Grantham, Newark, Retford, Doncaster, York, Darlington and Newcastle at 2s 6d each. If with a half bottle claret or burgundy, bottle beer or mineral water, at 3s 0d each. Any passenger desiring to have a basket should give notice to the guard of the train or to one of the station staff at a previous stopping station, in order that a telegram may be sent.

In order to prevent wrong delivery, passengers are advised to give their names to the official taking the order . . .[18]

Ardvreck School was founded in 1883 by Mr W. E. Frost, who was its first Headmaster.[19] The school is situated on a hillside on the outskirts of Crieff and was built with local red sandstone from the nearby village of Dunning, as were many of the older houses in Crieff. The school house was similar in appearance to the country mansion of a Scottish laird; its solid construction, gables, curved fortress-like walls, turreted roof line and multiple chimneys were built to last and to withstand the cold and winds of wintertime. Over time, the walls became covered with creepers and climbing roses. The situation of the school was magnif-icent, overlooking the broad valley below, and with the Perthshire hills rising up at the back. In Wood's day, the immediate environment was rather open and bare, but many trees were planted at that time and now, one hundred years later, the school is surrounded by huge, mature trees and beautiful gardens full of rhododendrons. Mr Frost founded the school with the intention of teaching his pupils 'patience, justice, obedi-ence and unselfishness';[20] in addition, true to the best traditions of a preparatory school, the pupils led a full life of formal classes, sports, drama, clubs, and many extra-curricular activities. In its early days, Ardvreck was an all-boys school, and the school uniform was very formal: high buttoned jacket, white collar, necktie, long trousers and strong shoes. When representing the school in sports, boys wore blazers and sports shirts in the school colours of green and white. Hair was cut short and neatly parted.

When Wood first arrived at Ardvreck in May 1899, there were just fifty boys in four classes. He entered Form I and during his four years he progressed to Form VI. At first, it must have all seemed very strange and daunting: he was far from home, and living in an environment very different to what he was used to in London. The school magazine provides a wonderful snapshot of school life and accomplishments during the years 1899 to 1903. Outside their classes, the boys had lots of fun: Hallowe'en night was celebrated in the traditional manner, there were bonfires and fireworks on Guy Fawkes' Day, and there were regular paperchases. During the winter months boys indulged in rugby, skating, curling and tobogganing, and in summer there was cricket, squash, fives, fishing and numerous picnic excursions.[21]

There were many activities for boys interested in natural history. Mr Frost sometimes gave evening lectures, when he used magic lantern slides to show and describe life in different regions of the world. Especially memorable and enjoyable to pupils was his talk on 'The Dark Continent' using pictures taken by missionaries on the Congo River, and no doubt he encouraged boys to learn more about Africa and other foreign lands. The Natural History Society was thriving when Wood arrived, with twenty boys coming to the Tuesday evening get-togethers. Here, they talked about butterflies, examined their collections, and discussed all sorts of natural history topics. Collecting specimens was an integral part of natural history at this time. Mr Frost was a determined collector, and had shot many birds of different species which were mounted by a taxidermist and displayed in glass-fronted cabinets in the school library.[22] On Saturdays, during the warm summer months, there were 'Barvicks' – so named after Barvick Glen nearby – when boys who were interested in natural history had expeditions into the surrounding countryside.[23] Accompanied by a member of staff, the boys spent a day or an afternoon walking and swimming, and collecting insects and flowers and anything else that took their fancy. An appreciation of outdoor life and local natural history was part of the ethos of Ardvreck: each issue of the school magazine contained notes on local flowers, butterflies, the nesting of birds, local fishing records and many other items of interest to the young naturalists.

Wood participated in many varied activities at Ardvreck. In 1901, he was one of the fifteen members of the 'Glee Club' and participated in the Christmas concert. The description of the concert was graphically described in the school magazine as follows:

> On the last Saturday of the Xmas term, the Glee Club gave us a concert. A varied programme of glees and piano solos was well rendered, many of the items being loudly encored. The glees, which were all pretty and tuneful, were sung with plenty of 'go', and showed that the members of the Glee Club had taken a keen interest in the practices . . . The programme was divided into two parts, with an interval, during which refreshments were served . . . The concert was concluded with 'God Save the King', the first occasion, we believe, on which the new version of the National Anthem has been sung at Ardvreck. The glees included 'Who will over the downs with me', 'April showers', 'Boat song', 'Valet' and 'The Old Brigade'.[24]

Wood was still a member of the Glee Club when it presented a 'Nigger Minstrel Entertainment' at the Christmas concert in 1902.[25] The school magazine provides another graphic description:

A varied programme of solos and choruses was much enjoyed by the audience, among which were several visitors. The performers looked quite at their best with black faces, shown up by white collars and large blue and pink bows. The blacking of fourteen faces, preparatory to the concert, was no small undertaking, but the effect quite compensated for the trouble . . .

Wood's love of singing and music, which he retained throughout his life, was encouraged at Ardvreck. All these songs and Negro spirituals are forgotten or rarely heard now, but they attest to the fun and enjoyment of school concerts in those days.

Besides the Glee Club, Wood participated in various other activities at Ardvreck. He was a keen member of the Natural History Society, and 'had the best collection (of butterflies and moths) in the school'. In 1902, he won the Drawing Prize. At the Halloween Party of the same year, he was one of one of the six best dart players, and although he came second to the winner, he was the only competitor to score a bull's eye. During his last year at the school, 1902–1903, he was Senior Boy, a member of the school cricket XI, and the Form VI prize winner for the Christmas and Midsummer examinations. In early 1903 he passed the entrance examination for Harrow. The school recorded this achievement with a certain hint of pride: 'Wood, we hear, did very well indeed and is likely to take a high place at entrance.'[26] Ardvreck School had served him very well, as it has all its pupils, enabling many of them to enter the top public schools throughout Britain. As one pupil stated in 1998, 'Ardvreck provides a secure start to life', just as it did in Wood's time. Wood did not forget his prep school, and in the first few years after leaving, he wrote to Ardvreck about his life and progress at Harrow. A former pupil cannot express greater appreciation than this!

At the end of the summer term of 1903, Wood left Ardvreck and returned to London, and in September 1903, he enrolled as a new boy at Harrow School.[27] Harrow is one of the oldest and best known of the English public schools. Many of its old boys, later in life, have become famous and have had considerable influence on the society of their times. By sending Rodney to Harrow, Alexander Wood was hoping for the best possible education for his son. The school is located in the little town of

Harrow-on-the-Hill, some fifteen miles north-west of London. In fact, virtually all the town is the school. Most of the school buildings line the High Street, together with a few little shops and private houses. At the beginning of the twentieth century, when Wood first went there, farms, meadows and fields separated Harrow-on-the-Hill from London.

On arrival at Harrow, Wood lived in a small boarding house called West Hill, but after his first term he moved to Druries for the remainder of his school career.[28] Druries is a large, stern, and very Victorian building, made of red brick. It has three floors and its roofline, like so many buildings of this period, is punctuated by numerous chimneys. On one side is a large garden on a steep slope with many large beech and sycamore trees, and beyond are the extensive school playing fields. In Wood's day, Druries was not visible from the High Street because there was a row of little houses and shops between it and the street. (These houses were removed in the 1920s to allow for widening of the street.) His housemaster during his first two years, Mr Howson, appears to have been a very remarkable man, 'intellectual and athletic, a fearless rider, an eager traveller and an accomplished writer of English verse'.[29] After Mr Howson's death in late 1905, Mr Stephens became the new Housemaster of Druries. Both men would have had a strong influence on Wood's activities, as well as on his progress and happiness at Druries.

Harrow has many unique idiosyncrasies. Quite logically, terms were referred to by their position in the year – for example, 1904^1, 1904^2 and 1904^3 – whereas other schools would refer to spring, summer and autumn terms, or to Lent, Summer, and Michaelmas terms. Likewise, members of a house team were not referred to by their house name, but by the name of their housemaster – 'Mr Stephen's' or 'Mr Howson's'. The naming of the various year-classes is also somewhat confusing to the uninitiated. A new boy was normally placed in 'Shell', which comprised several forms which were streamed on merit. He then progressed to 'Removes' and 'Lower Fifth' (both classes in 'Middle School'), and finally he reached the Upper School, comprising 'Upper Fifth' and 'Sixth Form'.[30]

The records for Wood's progress through the school are patchy. Nevertheless, his work was consistently good and in all years he remained in the top third of his class of about 30 pupils.[31] By 1906, he had progressed to the Sixth Form. During his time at Harrow, he was awarded the form prize for 'Modern Fifth Third Remove' in 1904[2], and a prize for natural science in 1906. Of particular interest is the record that he was commended to the headmaster four times for his work in natural science.[32] He did not participate in sports to any great extent, at least not to the level where he was in a school or house team, although his house handicaps suggest that he was quite good at running. Wood left Harrow at Christmas 1906 (that is, at the end of 1906[3]). It is difficult to understand why he left at this time. Most boys leave school at the end of the summer term and then, if university entrance has been attained, they have a couple of months holiday before beginning their university careers. Although some boys evidently did leave Harrow at the end of the Christmas term after obtaining a place at university, there is no record that Wood attempted, or gained, university entrance, even though his school record certainly suggests that he was bright enough to have continued his studies at university.

During Wood's schoolboy years, there were many accepted customs, social changes and important historical events that would have affected the way he was brought up. It was an era of formality and correctness, especially in dress. At both Ardvreck and Harrow there were strict regulations about uniform, and boys were expected to be suitably attired at all times. Formal dress at Harrow was morning dress with tails and top hat; during normal school times, boys wore dark long-sleeved jackets, white collars, ties and long, dark trousers, and, if outside, a boater with a coloured ribbon.[33] Wood's father appears to have been a very formal person, and no doubt Rodney also had to comply with accepted conventions of dress during the holidays. Even children playing in the parks and at the seaside at this period were formally dressed; the lax approach and freedom given to children today would have been unthinkable during Wood's childhood.

There were many major changes which affected day-to-day living at this time: electricity became more widespread and replaced gas and candles in homes; telephones were installed in businesses, schools and residences; and wireless sets (large heavy instruments full of valves, tubes and wires) brought news and entertainment into the home. The gramophone was also popular with many people; it had a huge trumpet-like speaker, and the turntable (revolving at 78 r.p.m) was wound up with a handle. Motorised transport gradually replaced horse-drawn carriages, and roads were improved and sealed with tarmac. Ocean liners and cargo boats, powered by engines, crammed the ports, and overseas travel (for those who could afford it) was by boat and train. Modern inventions allowed greater mobility, ease of communication, and a more interesting and pleasant lifestyle.

Wood's schoolboy years coincided with the expansion of the British Empire. Many explorers, administrators, missionaries, planters and traders went out to the colonies, and their stories and experiences must have been a source of inspiration to young boys at Aardvrek. Mr Frost's lectures at Ardvreck no doubt caught the imagination of many of his pupils and inspired them to pursue a career overseas, and Wood's first knowledge of Africa and its wonderful animal life, and his desire to go there, must have originated at Ardvreck. There is no record of how many of Wood's contemporaries at Ardvreck went overseas, but the school recorded with sadness that Mr Frost's brother and four Old Ardvreckians were killed in the Boer War.[34] The dangers and costs of administering the British Empire would have been very evident to the boys of this time, yet for many of them – such as Wood – the attraction of a career in the Empire was irresistible.

At both Ardvreck and Harrow, Wood became increasingly knowledgeable and curious about the natural world. Two of his teachers, in particular, had a strong and encouraging influence on the development of his talents – Mr Frost, his Headmaster at Ardvreck, and Mr Archer Vassall, his Biology teacher at Harrow. One cannot overestimate the importance of good teachers who have the ability to encourage students so that their natural talents, whatever they may be, are developed to the greatest extent. Mr Vassall was evidently one of that small band of brilliant housemasters and scientists who had this ability.[35] To what extent he taught Wood is not known; but Wood knew him sufficiently well that in 1928 he was able to suggest that Mr Vassall might write a reference for his application for the post of Game Warden in Nyasaland (see Chapter 6). And surely Mr Vassall was the motivating force behind this delightful letter that Wood wrote to the British Museum of Natural History in 1905:

Druries
Harrow
July 8th

To G. A. Boulenger Esq.

Dear Sir,

I enclose five sticklebacks which have all got four dorsal spines. I do not think they are the same fishes as the four-spined stickleback (Gasterosteus spinulosus), both on account of their spines which are 2 large ones and 2 short ones, and chiefly because of the size. Also the cocks had red throats when caught, which have now considerably faded in the spirit. All of these things are different in the four-spined Stickleback in Yarrell's Fishes; so on the recommendation of A. Vassall Esq, Biology Master here at Harrow School, I have sent these sticklebacks to you, hoping that you can give me some information about them. Please return them as soon as possible, for which I enclose 6d in stamps, for postage.

    I remain
    yours truly
    Rodney C. Wood

P.S. The fish were caught in a pond here on July 6th. There were about 1 of these in every ten fish I caught. The others were all 3-spined sticklebacks.[36]

The detail in this letter shows that, even when he was only 16 years of age, Wood was a careful observer, was familiar with scientific names, and could record essential information necessary for identification. And what a thoughtful touch to enclose 6d for the return postage!

During his formal education, Wood developed many characteristics and abilities which were evident throughout the rest of his life. He learned to be very observant, to be meticulous in recording information (whether for collections, notes or accounts), and to be punctual and well-mannered. He wrote good prose with beautiful handwriting, was a prolific letter writer and a capable diarist, and could tell many stories of his adventures and experiences. He developed a love of books, especially reference books and those which provided information. He had a good eye (as indicated by his success in darts), which allowed him to be a good hunter and marksman when he first lived in Africa. Many people who knew him later in his life commented that he had a wonderful sense of fun and a wry sense of humour, and that he spoke with a 'well-bred English accent'. He loved to talk, to tell interested listeners about his collections, and to discuss and converse on a wide range of topics. The earliest photograph of Wood that has survived was taken when he entered Harrow School in 1903 when he was nearly 14 years of age. It shows him to be slimly built with a rather aquiline-shaped face, wide mouth, rather large ears and well-groomed hair – and perhaps with a touch of mischievousness and diffidence about him. Another photograph, probably taken a few years later, shows Wood with his father. Edwardian portrait photographs were always very formal. Both father and son are dressed in dark suits, with high white collars and ties. Father is wearing a waistcoat and holds a cigarette in a cigarette holder, and son holds a pair of gloves and a walking cane, with his boater hat resting on a chair; behind, on a backdrop, there is the misty form of a country house, set in parkland, with a lake and swirling clouds. There is a striking resemblance between father and son, especially in the shape of eyes, mouth and nose. Physically, Wood developed into a strong well-proportioned young man; tall – he was six feet and half an inch by his own reckoning – and slim, with soft brown hair and bright, penetrating blue eyes.

Exactly what Wood did in the years 1907, 1908 and 1909 after leaving Harrow are unclear. He recorded that he was 'travelling in France learning the language'.[37] Many wines were imported from France at the beginning of the twentieth century, and learning French was a useful, if not essential, qualification for good business. Later in life, Wood was able to draw on his knowledge of French when he taught at a school in South Africa, and at times he conversed in French when living on the Seychelles. It is highly likely that Alexander Wood wanted his son to enter the wine trade and, perhaps, to take over the firm of Alex. Wood and Campbell in later years. In all probability, Wood was associated with his father's wine business for some of these years, for on one occasion he travelled with his father to Shanghai, presumably on business. To become a vintner required considerable training, and a normal procedure was to be apprenticed to a firm for a number of years. In an ancient leatherbound book listing the Apprentices 1891–1908 of the Vintner's Company, there is the following entry:

Feb 13th 1908
Rodney Carrington Wood son of Alexander John Wood, Wine Shipper of 36 Abercorn Place, NW to be apprenticed to Robert Gray Citizen and Vintner in consideration of service for 4 years.[38]

This means that after four years of apprenticeship with Robert Gray, Wood was entitled to be a vintner in his own right and to become a Freeman of the Vintner's Company. However, it seems that Wood did not want to pursue a career in the wine business, because at some point in 1909, he travelled to Southern Rhodesia to work on a farm. He never returned to the wine business. Yet, on 2 April 1914, the records of the 'Vintner's Freedoms' has the following entry:

Rodney Carrington Wood, P. Servitude[39]

Thus, Wood was admitted to the Freedom of the Vintner's Company by servitude; that is, he had served his four years of apprenticeship. (The alternative entry, 'P. Patrimony', recorded for other Freedoms given by the Company, is when a son follows his father as a member of the Company.) But Wood cannot have completed his four years with Robert Gray, because at the time he was due to complete his apprenticeship in 1912, he had already been in Southern and Northern Rhodesia for at least two years, and by 1914 he was in Nyasaland. How he managed to gain his Freedom must forever remain a mystery!

So, in 1909,[40] when he was 20 years of age, Wood turned his back on the formality of England and on the world of business. All his training, especially in natural sciences, equipped him for a life in Africa. He remained there and in the Seychelles for the rest of his life, returning to England on only one occasion in the early 1920s. Once in Africa, he started making his collections and observations, and it is for these that he will long be remembered.

Chapter 2

LIVINGSTONE'S LAND

Most of Wood's life was spent in Nyasaland, a small land-locked country in eastern central Africa. Nyasaland first came to the notice of Europeans because of David Livingstone's travels and his efforts to abolish the slave trade. At that time, the 1860s and 1870s, this part of Africa had no proper name; it was part of the vast unknown African continent where there were no national boundaries and few maps. In 1891, Lake Nyasa and much of the surrounding hills and rivers became known as the 'Protectorate of British Central Africa', a name that was changed to 'Nyasaland' in 1907 and to 'Malaŵi' in 1965. The country is still full of legacies of David Livingstone and the missionaries who followed him, with place names such as Livingstonia, Blantyre, the Livingstonia Mountains (just over the border in Tanzania), the Kirk Range, and Cape Maclear recalling the early days. There are graves and other places of historical interest that are further reminders of the lives and endeavours of Livingstone and his contemporaries. Even Livingstone's original description of the lake, 'The Lake of Stars', describing the glittering reflections on the surface of the rippling waters, is still used to evoke the beauty and magic of Lake Nyasa. Other epithets have been coined to describe the country: 'The Land of the Lake' recognises the importance and magnificence of Lake Nyasa, and 'The Warm Heart of Africa' describes the friendliness of its inhabitants. Much of the early history of Nyasaland is intimately connected with the endeavours of missionaries who came out from Britain. They established missions in remote parts of the country where they set up schools, craft shops, hospitals and churches. Their zeal and determination is legendary. The Universities Mission to Central Africa (UMCA) was the first Protestant mission to settle in Nyasaland, originally at Magomero (near Zomba), and later at Likoma Island. The

Free Church of Scotland established 'Livingstonia', which was located at Cape Maclear but subsequently moved further north to Bandawe and finally Kondowe. And the Established Church of Scotland settled among the hills of the Shire Highlands, and called its mission Blantyre, after the Scottish town where Livingstone was born. Wood had the good fortune to live in many places of historical interest in Nyasaland, so it is appropriate to begin this story of his life with an introduction to the country and the places – Chiromo, Cholo, Cape Maclear and Monkey Bay, Livingstonia, Blantyre and Zomba – where he lived and worked.

The Landscapes of Nyasaland

Nyasaland is a small country by African standards; it is situated in east central Africa [between 09°S and 17°S, and between 33°E and 36° E] and is bordered by Tanzania, Mozambique, and Zambia. It is a long narrow country, some 900 kilometres long and 160 kilometres at its widest point, and it has an area of about 118,500 square kilometres. The lake, one of the major features of the country, is 580 kilometres long and 16–83 kilometres wide, and covers 23,000 square kilometres. Most of the lake is within the jurisdiction of Nyasaland, but parts of the eastern shore and inshore waters belong to Mozambique or Tanzania.

The landscapes of Nyasaland, as we see them today, are a result of a long geological evolution. Two hundred and fifty million years ago, Africa was part of the great southern continent of Gondwanaland and was situated much further south than it is now. This southern continent also included South America, India, Australia and Antarctica. When Gondwanaland broke up about 160–170 million years ago, Africa separated from the other continents and drifted slowly northwards until it came to lie astride the Equator. Its animals and plants were isolated on the African continent and were able to evolve into the uniquely 'African' animals and plants that we know today. However, they still retain some similarities and relationships with the animals and plants of the other continents that formed Gondwanaland – a fact which helps us to trace the past history of each of these continents.

The geological evolution of the African continent has been extremely complex. There have been successive waves of uplifting, faulting, warping, rifting, erosion and sedimentation which have produced the complex landforms that we know today. The huge uplifted plateau which forms the majority of southern and eastern Africa is composed of ancient rocks of pre-Cambrian age, laid down more than 600 million years ago. Over many millions of years, warping, faulting and erosion have exposed rocks of different ages on the surface, resulting in a diversity of rock patterns throughout the plateau. The original plateau is divided now into several separate plateaux; on the edge of each plateau, the hills and valleys gradually decline in elevation until they reach the large rivers which ultimately flow into the sea. One of these plateaux, the Central African Plateau, is bordered by the Luangwa valley in Zambia in the west, the Zambesi River in the south, and the lower elevations of Mozambique in the east. During the Miocene period (about 10–20 million years ago), a major geological disturbance caused a huge fault line to develop along the length of eastern Africa. This resulted in the formation of the Rift Valley, which stretches from the Red Sea in the north, through the middle of the Ethiopian plateau and East Africa, to Nyasaland in the south. The valley was formed by two distinct but related processes: the uplifting of the sides of valley, and the subsidence of the valley floor. The Rift Valley is characterised by the large, deep lakes which have formed in the valley, and by the steep escarpments on its sides. In East Africa, the Rift Valley divides into two: the western Rift, which runs through Uganda, Zaire, Tanzania and Malaŵi; and the eastern Rift, through Kenya and Tanzania. In Malaŵi, the Rift Valley slices right through the country in a north–south direction; Lake Nyasa is situated in the northern half, and the Shire River (pronounced "Sheer-ray"), which drains the lake, is in the southern half. Escarpments border the length of the valley; these may be high and steeply sloping, or only a series of low hills. The influence of geology and climate is very obvious in Malaŵi: the landscapes are very varied and beautiful, and there is a wonderful diversity of vegetation. At one extreme are the high cool mountains and plateaux rising up to 3,000 metres, and at the other are the low, hot savannas close to the Zambezi valley. There are six main landscapes in Nyasaland: the Central African Plateau; the high mountains and plateaux; the Rift Valley escarpments; the lake; the lakeshore and surrounding plains; and the Lower Shire Valley.

The Central African Plateau in Nyasaland is part of the extensive plateau country of southern Africa. The plateau is relatively flat, with steep hills and valleys in places, and bisected by small meandering streams and rivers. Where there has been extensive erosion over time, rocky inselbergs emerge from the surrounding plains. The plateau is covered by savanna woodland, with marshes (*dambos*) in low-lying areas. The plateau extends along the western side of the lake and along both sides of the upper Shire River. The climate on the plateau is very pleasant because of its altitude of about 1,000–1,500 metres above sea level. In the warmest months of the year, the daily temperature is around 15–30°C, and in the cool season it is 8-23°C. In addition, the alternating wet and dry seasons provide a wonderful contrast in the vegetation: during the wet season (December to April), the savanna grasses are tall and green, the trees are thick with leaves and there are many flowers; in the dry season (May to November), the grasses die and turn to a golden oatmeal colour; many grasses are burned to the ground so the countryside has a wide open feeling, and some species of trees lose their leaves. The cycles of wet and dry, and of hot and cold, result in a colourful pageant of colours and textures throughout the year.

In a few places, isolated mountains and plateaux rise above the Central African Plateau. In the north of the country, the Nyika Plateau (1,500–2,500 metres high) is an extensive area of rolling hills covered with short tussocky grasses and heaths, with patches of moist evergreen montane forest in gullies and on protected slopes. Many fast-flowing streams of cool clear water run through this region. Similar, but slightly lower, is the Vipya Plateau (1,500–2,000 metres high), but in recent years much of this plateau has been afforested with pine plantations. In the far north, the Mafinga Mountains (which extend over the border into Zambia) and the small Misuku Hills (close to the Tanzanian border) are also covered with grasslands and evergreen forests. In the south of the country, Zomba Plateau, which borders the eastern rift escarpment, is a small area of montane grassland and evergreen forest, and nearby is Mulanje Mountain, a huge rocky massif with precipitous sides which rise almost 2,000 metres above the surrounding countryside. Because of its height, the mountain has many vegetation zones – a walk up it takes one, within a few hours, from the savanna woodlands at its base through various forms of evergreen forest and heathland to the montane grasslands and forested valleys below the summit. Above the grasslands, the summit (at 3,002 metres, the highest point in Nyasaland) is bare rock and boulders mostly devoid of vegetation. The climate of these mountains and plateaux is cooler and wetter than that on the Central African Plateau, and frosts are not uncommon during the cool season. The mountains and plateaux are of great interest to biologists because their specialised flora and fauna provide clues to their biogeographical relationships with the other mountains and high plateaux of eastern Africa.

The Rift Valley escarpments link the Central African Plateau with the lakeshore and the Shire Valley. In places, the escarpment is rocky and precipitous; in others, it is a series of steep hills and gullies. The views from the top of the escarpment are magnificent; at Vipya, for example, the land falls from about 1,500 metres above sea level to 480 metres at the lakeshore, over a distance of only a few kilometres. The hills and the lakeshore are a pattern of various greens and greys – the colours of forests, small cultivated fields and huts. Beyond the golden line of the shore, the lake is a pageant of blues and greys; clouds drift across the sky, their shadows forming dark patches on the glinting waters. If it is windy, white crests of waves can be seen scudding across the surface of the waters. Far away in the distance, the Livingstone Mountains in Tanzania may be visible through the mist and haze. Above all is the immensity of the sky, sometimes blue, sometimes full of little white clouds, and in the wet season, grey and thundery. Similarly, looking west-wards and downwards from Zomba Plateau, the wide valley of the Shire River appears in the wet season as a mixture of many shades of green (or grey and brown in the dry season); the river is a thin silver snake-like line, and on the far horizon is the Kirk Range, misty greyish-blue in colour and usually topped by banks of clouds. The western escarp-ment stretches from the far north southwards to the Kirk Ranges, a distance of about 700 kilometres. The eastern escarpment is not so well defined. Although it is well developed in the Livingstone Mountains (in Tanzania) which border the north-eastern edge of the lake, it is less obvious at the southern end of the lake. The escarpment becomes well developed again to the east of the Shire Valley, where the Shire Highlands and Cholo Hills rise some 1,000 metres above the valley floor. The vege-tation and cool climate of the plateau country at the top of the escarp-ment provide a very stark contrast to the dry savanna and hot climate of the lake and the lakeshore. There are many stunning waterfalls on the escarpments where the rivers on the plateaux descend to the lake or the Shire River, all of them rich with ferns, mosses and luxurious vegetation.

Lake Nyasa is one of the most beautiful of the freshwater lakes of Africa. The water is clear and deep, there are long sandy beaches interspersed with rocky headlands, high mountains line its shores, and it is full of fascinating fish and other wildlife. Normally a gentle breeze ripples its surface, throwing up little white-crested waves which glint in the sunlight. However, at times the lake is whipped by high winds and heavy rains; when this happens, it is too dangerous for canoes and sailing boats to set out for fishing expeditions.

The shape, size and water level of the lake has changed dramatically during the course of geological time. It is thought that the lake formed about two million years ago, rather late in the evolution of the Rift Valley. At first, it was small and situated somewhere near the northern end of the present lake. It was also some 200 metres above its present level because the Rift Valley was still in the process of rifting and warping. Deepening of the floor of the valley allowed the lake to increase in size and depth. However, a huge ridge of old rock runs across the valley at the southern end of the lake, which maintains the present level of the water at about 474 metres above sea level. Lake Nyasa is very deep; at its deepest point near Nkhata Bay its depth is 704 metres (of which 230 metres is below sea level). Because of the steep hills surrounding the lake, its catchment is small in relation to its area. The lake is justly famous for its huge variety of fish, and for its high productivity. Because of its vast size, it is more like a sea. Tall waves develop in windy weather, and the waters can then be very treacherous for boats, both big and small. There is only one outlet from the lake – the Shire River in the south. Because of the large volume of the lake and its small catchment, it is calculated that it would take about 700 years of average rainfall on the catchment to fill the lake to its present level.

The water level of the lake is always changing. Quite apart from the various changes over the course of geological time, there are seasonal fluctuations in level that have been recorded since Livingstone first saw the lake in September 1859. The lake level varies annually by as much as 2 to 2.5 metres; it is at its highest at the end of the wet season (April to May), and at its lowest at the end of the dry season (November to

December). But there are also longer-term fluctuations. Livingstone saw the lake for the first time at one of its high periods; from then to about 1915, the average level declined by about 4.5 metres, and then rose again by about 6 metres between 1915 and 1960. These annual and longer-term fluctuations in water-level in the lake have a dramatic effect on the water flow in the Shire River.

The lakeshore and the plains of the Upper Shire Valley are characterised by a much hotter climate than that on the plateaux. The lakeshore is a narrow strip of land up to 25 kilometres wide, although in some places where the escarpment touches the lake it is almost non-existent. Many of the soils are alluvial and rich, the result of soil erosion on the steep slopes over many years. The vegetation is quite different to that on the plateau; here there are different sorts of trees – acacias, combretums and mopanes, as well as many baobabs. Where large rivers emerge from the escarpment, there are *dambos* (swamps), and much of the shore is strewn with large boulders, some of which were originally higher up the escarpment. The climate at the lakeshore is warmer and drier than that on the plateau because of the lower altitude; the typical daily range of temperature is 22–29°C in the hottest season of the year, and 15–24°C during the coldest season. The Upper Shire Valley extends for about 130 kilometres south of the lake and is topographically part of the lakeshore; the altitude, climate and vegetation are similar to those of the lakeshore.

The Lower Shire Valley is a southern extension of the Upper Shire Valley, and is at a lower altitude (100 metres above sea level). The Shire River reaches the lower valley by flowing over a series of twelve cataracts (collectively named the Murchison Falls) which are spread over about 60 kilometres of the river. The cataracts prevent the passage of any boat between the upper and lower valleys, and ever since Livingstone's time, they have been a major obstacle to transport on the river. Soon after the river reaches the Lower Shire, it fans out through a huge marsh called the Elephant Marsh. The water level and water flow in the marsh, and along the length of the Lower Shire, fluctuates annually and from year to year. These fluctuations are due to the changing water levels in the lake, described above, and to variations in local annual rainfall. The many problems of navigation experienced by Livingstone's expedition and by subsequent travellers on the Shire River were due to these variations in water level and water flow. Apart from the Shire, all the rivers flowing from the hills bordering the Lower Shire are seasonal, flowing only during the wet season. Thus, the soils are dry and parched in the dry season, and soggy and flooded in the wet season. The alluvial soils of the valley are fertile, and are now used for growing sugar cane and cotton. The natural vegetation is a mixture of low-altitude wooded savanna and dense thicket-scrub. The Shire River leaves the southern border of Nyasaland through another marsh, the Ndinde Marsh, and flows on to join the Zambesi River. The Lower Shire experiences the highest temperatures of anywhere in Nyasaland, and in the wet season when the humidity is high, the climate can be unpleasant and enervating.[1]

Brief Histories of the Places where Rodney Wood Lived

CHIROMO AND THE LOWER SHIRE

Chiromo, the first place in Nyasaland where Wood lived, is steeped in history and was the site of many major events in the early history of Nyasaland. When the first European travellers steamed northwards up the Shire River they reached a point, some 140 miles north of the junction with the Zambesi River, where the Ruo River enters the Shire River from the east. The Ruo rises on the high rocks and plateaux of Mulanje Mountain; as it descends into the Rift Valley, its waters fall over spectacular rapids and then flow swiftly towards the Shire River. On the northern side of the Ruo, where it flows into the Shire, there is a low spit of land where, in the early years of the Protectorate, the little town of Chiromo was established. To the north of the junction is an extensive marsh, where the Shire River spreads out into numerous channels separated by soft marshy country thick with reeds. The marsh is about 25 miles long by seven miles wide at its widest point; however, the main channel of the Shire meanders through the marsh, so its navigable length is about 35 miles. To the early explorers, the marsh was a nightmare because their boats were constantly grounded in the shallow water. During the dry season, it was barely possible to travel through the marsh because the water was so low; in the wet season, the channels were deeper, but the constantly changing water flow and deposition of silt made navigation difficult and uncertain as old channels became blocked and new channels opened. The marsh was home to many elephants, and other large game, and was given the name of 'Elephant Marsh', the name that is still used today. The journey to Chibisa or Katunga from Chinde, at the mouth of the Zambesi, was long and wearisome: 160 miles to the junction of the Shire, 140 miles from the junction to Chiromo, and finally 50 miles through the marsh to Chibisa (or Chibisa's, so named after the village chief), or Katunga; a total of 350 miles of river travel. A few miles north of Katunga are the first of the cataracts which make it impossible for steamships to travel any further north-wards.

Livingstone and his party first reached the junction of the Ruo and Shire rivers in January 1859.[2] This region of southern Nyasaland was populated by the Manganja tribe, and several villages belonging to these tribesmen were situated along either side of the Shire River during the years of Livingstone's expedition. Livingstone and his colleagues passed the Ruo junction on numerous occasions as they made their way to or from Chibisa's village at the northern end of the marsh. From Chibisa's they climbed up to the highlands where, in later years, the towns of Blantyre and Zomba were established. The junction of the Shire and Ruo rivers (where the little town of Chiromo would be established in later years) was a good meeting point, and a good place to stop at the southern end of the marsh. It was also the scene of one of the most tragic episodes of missionary history in central Africa.

In May 1861, Bishop Charles Mackenzie, the first Bishop of Central Africa, arrived at the junction of the rivers. He had been Archdeacon of Pietermaritzburg for some years, so he was no stranger to Africa when he accepted this new appointment. The Shire river was very low when he arrived and the ship he had travelled in, the *Pioneer*, became stranded on sandbanks on many occasions along the way. It took 24 days to cover just twelve miles in one part of Elephant Marsh, and the bishop and his companions eventually reached Chibisa's on 8 July, arriving at Magomero on the Shire Highlands (where the first UMCA mission was to be established) on 19 July.[3] Such a long and arduous trip was not unusual when the river was low.

Six months later, in January 1862, Anne Mackenzie (Bishop Mackenzie's sister) and Mrs Burrup (the wife of the Reverend Henry Burrup) were due to join the mission at Magomero. It was arranged that the two ladies would come upstream in the *Pioneer*, and would meet the bishop and the Reverend Burrup at the Ruo junction on 1 January 1892. Mackenzie and Burrup were delayed (partly because Burrup was ill) and they did not reach the junction until 11 January, where they were expecting the ladies to be waiting for them. They asked members of the local Manganja tribe if they had seen the *Pioneer*, and were told that the boat (with David Livingstone, Charles Livingstone, John Kirk and Dr Meller on board) had travelled downstream four days previously;[4] it transpired that the *Pioneer* had been badly delayed in the Elephant Marsh on its way to the coast to meet Miss Mackenzie and Mrs Burrup, and was a long way behind schedule. Mackenzie and Burrup decided to keep to the original plan, and arranged with the local chief to stay in a hut on Malo island close to the junction to await the return of the *Pioneer*. After a few days, both men suffered from diarrhoea and fever. Burrup nursed Mackenzie, but the bishop did not respond to treatment and he died on 31 January 1862. Mackenzie's body was taken to the mainland and buried under an acacia tree on the east bank of river just south of the junction.[5] Burrup was so ill that he insisted on returning to the cooler climate at Magomero in the highlands; he left a letter for Livingstone explaining what had happened. It is difficult to imagine the pain and hardship he must have suffered on his journey. The Shire was now so swollen from the heavy rains upstream that paddling the canoe against the current was impossible; so he staggered along the bank for many days, and collapsed when he reached Chibisa's. The three Makokolo who accompanied him carried him up the escarpment, and they arrived at Magomero on 14 February. Burrup's condition deteriorated, and he died on 22 February. He was buried in the grounds of the mission at Magomero.[6]

Meanwhile, the *Pioneer* was steaming down the Zambesi River towards the coast, and Livingstone and Kirk were unaware of the tragedy on Malo Island. The *Pioneer* arrived at the Luabo mouth of the Zambesi on 30 January 1862,[7] and on 3 February Livingstone and Kirk met the ladies who had travelled from Durban, via Mozambique and Quilimane. It took several days to load the masses of food and supplies that had arrived from England and South Africa and which were necessary to provision Livingstone's expedition. In addition, there were sections of the new boat, the *Lady Nyassa*, which Livingstone intended to assemble on the lake, as well as clothes, furniture, agricultural implements, cooking utensils, photographic apparatus, thousands of fish hooks, a 'church tent', two farm carts, two mules and Katie, the donkey who was to carry Anne Mackenzie because she was unable to walk any distance.[8] On board were David Livingstone and John Kirk, Ann Mackenzie, Mrs Burrup, Mary Livingstone (David Livingstone's wife), Jessie Lennox (the housekeeper), Sarah (Anne Mackenzie's maid), and James Stewart (a Free Church minister who was also a medical student). Besides the *Pioneer*, there was another boat, HMS *Gorgon*, which was also laden with stores, passengers and sailors (in those days called 'blue-jackets'). The flotilla was led by Commander Wilson, a naval officer, with a hundred sailors under his command to man the boats.[9] It is difficult to imagine how this ill-assorted mixture of people managed to fit themselves and their stores onto the small boats. Nevertheless, the flotilla left the mouth of the Zambesi and steamed upstream – not without problems – to Shupanga (a village on the Zambesi River downstream from the Zambesi–Shire junction). Here, Captain Wilson decided to push ahead with the gig (small boat) from the *Gorgon*, with Ann Mackenzie and Mrs Burrup, Ramsey (the *Gorgon*'s doctor) and eleven blue-jackets on board. John Kirk and Sewell (a naval lieutenant) also set off in a whaler from the *Gorgon*.

They left Shupanga on 17 February; by now, they were six weeks late for their meeting with the bishop and the Reverend Burrup, and they were still some 200 miles from the Ruo junction. Like so many travellers on these river journeys, they suffered from the heat and exposure, and on one occasion they lost their way in the river channels at the entrance to the Shire. Miss Mackenzie became seriously ill, and by the time they reached the Ruo junction on 28 February, she was lying prostrate in the bottom of the gig.[10] Of course, they did not meet the bishop and the Rev Burrup, and had no idea that they had passed close to the bishop's grave. The local villagers did not inform them of his death. At the junction, Kirk transferred to the gig and Sewell returned in the whaler to Shupanga. Wilson, Kirk and their party, including the two ladies, made their way through Elephant Marsh, now full of water, to Chibisa's, where they arrived on 4 March and heard the devastating news that Bishop Mackenzie was dead and that Burrup had returned to Magomero. Miss Mackenzie and Mrs Burrup stayed at Chibisa's and were cared for by Dr Ramsey, while Wilson and Kirk, with many porters, walked to Magomero on the highlands with stores for the mission. At Magomero, they heard of the death of the Reverend Burrup. There was no now reason for Anne Mackenzie and Mrs Burrup to remain in Africa. They left Chibisa's and returned to Shupanga, briefly stopping at the Ruo junction, where Kirk and Wilson were shown the bishop's grave; Miss Mackenzie, however, was not well enough to accompany them to it. On 4 April 1862, all the ladies, including Jessie Lennox and Sarah, the maid who had been left at Shupanga, sailed in the *Gorgon* for Cape Town, never to return to Africa.[11] So ended one of the saddest episodes of the Livingstone expedition. There were many other stories of sadness, hardship, personal quarrels, and triumphs during these years, but Bishop Mackenzie's death at the Ruo junction was one of the greatest setbacks for the establishment of a mission in the highlands and for Livingstone's attempt to stop the slave trade in this part of Africa.

After a lull following the departure of Livingstone's expedition in 1864, the river traffic which passed the Ruo junction increased greatly. In the 1870s and 1880s, the town of Chiromo that would later develop at the junction had not yet been established, although there was a village inhabited by the Makokolo who had settled there after accompanying Livingstone on his various expeditions. Virtually all the people who became well known for their contributions to the growth of Nyasaland travelled up the Shire and passed the Ruo junction, as did all the equipment and stores necessary for development around the lake. Two years after Livingstone's death in 1873, when there was enormous zeal to continue his work, Dr Robert Laws of the United Free Church of Scotland, accompanied by Lieutenant Young and several artisans, steamed passed the Ruo junction in the mission boat *Ilala*.[12] The *Ilala* was brought out from England in sections and assembled at the Kongoni mouth of the Zambesi, and then travelled up the Shire to the Murchison Falls. At the Ruo junction, Dr Laws erected an iron cross (given to him by Anne Mackenzie) at the grave of Bishop Mackenzie.[13] Below the Falls, the *Ilala* was dismantled and packaged into 50-pound loads. These were carried by hundreds of porters for sixty miles, up a small path (probably the same path that Livingstone used unsuccessfully to carry the *Lady Nyassa* in 1860) to the top of the cataracts. Not a single item was lost! Lieutenant Young recalled his impressions of this remarkable feat as follows:

The best answers [sic] to our fears was the sight of the *Ilala* resting at her moorings, tight as a bottle and fit for anything. Yes, let this ever stand to the African credit; that 800 of these men worked, and worked desperately hard for us, free as air to come and go as they pleased, over a road which furnished at almost every yard an excuse for an accident, or a hiding place for a thief or a deserter, and yet at the end of sixty miles we had everything delivered up to us unmolested, untampered with, and unhurt, and every man merry and content with his well-earned wages of unbleached calico.[14]

The *Ilala* was reassembled and continued her journey up the Shire River to Lake Nyasa, where she arrived on 12 October 1875; she finally ended up at Cape Maclear, where the first Livingstonia mission was established.[15]

Many other people passed the Ruo junction during these years. Some are remembered because of their writings; others are remembered by nothing more than a name or a memorial in the cemetery. Some survived into old age; others died of malaria, other diseases or misadventure within a few months or years. Analysis of the fates of the 45 British missionaries who joined the Livingstonia Mission from 1875 to 1890[16] provides a general picture of the difficulties and hazards of life: eighteen of them died, eight were invalided and could no longer stay in Africa, and ten resigned or were recalled. Thus, over half died or had to leave because of ill health. It is also of interest that the five African missionaries from Lovedale in South Africa, where the climate and diseases are quite different to those of central Africa, did not fare any better than the Europeans: two died, two were invalided, and one resigned.

There was only a village, known as Chiromo or Tchiromo, at the Ruo junction in the late 1870s and 1880s. It became notorious because of skirmishes between chiefs and some of the European traders as a result of excessive drinking and unfair trading.[17] Few travellers stopped there; most went further north through Elephant Marsh to Chibisa's (or Katunga's), from where they continued by foot to the Shire Highlands or to Matope above the cataracts and hence to the lake. Among the notables who passed Chiromo were doctors and missionaries of the Livingstonia mission, such as Dr James Stewart, Dr Black, the Reverend

Walter Elmslie and his wife, and the Reverend Kerr Cross and his wife. There were many hunters and traders who sought adventure and profit mainly by shooting elephants and selling the ivory. Alfred Sharpe (who later became the second Commissioner of British Central Africa) first travelled up the Shire in 1887 in order to shoot elephants near the northern end of Lake Nyasa.[18] Captain (later Lord) Lugard, better known in later years as the Governor of Nigeria, had his first introduction to Africa when he steamed up the Shire in May 1888 on his way to participate in the 'war' against the slaver Mlozi at Karonga.[19] Then there was John Buchanan, one of the first vice-consuls and a successful coffee planter, who built 'The Residency' at Zomba. Two brothers who had considerable influence on trade were John and Frederick Moir, who founded the Livingstone Central Africa Company in 1878.[20] They steamed up the Shire in March 1879 bringing a new boat, the *Lady Nyasa* (named after Livingstone's first boat), with them. Frederick Moir recorded the problems of river travel in these early days:

The exceptional flood had spread over the valley for miles, destroying many villages, as was evident from huts and roofs floating past us out to sea. It had also carried away the stores of firewood prepared for the steamer, so that those on board had to stop at forests or villages on high ground and buy or cut down trees for fuel. Seldom could good dry trees be procured, so that raising steam was difficult and progress slow, and the constant cutting and sawing was hard on the crew. . . . The Shire wound its way with constant S-bends through this muddy morass, extending to hundreds of square miles. . . The flood had floated off quantities of sudd from the higher reaches and lagoons, which came down in floating islands. As the river was overflowing its mud banks, the current at the S-bends was often sluggish, and acres of dense masses of sudd collected and blocked the river while the water flowed in its deep channel below the floating masses. . . . Some of the first blocks encountered were removed by putting out grapnels into the thick mass at carefully selected spots, and with full steam downstream starting an enormous island down the current. The steamer then drew into the side, while the huge obstruction passed down. . . . Native fishermen in light canoes were met on the higher reaches, and they supplied fish and information as to channels. They also brought firewood from the higher land on the north. . . . As the waters fell to a more normal level, the increased current in the main river channel cleared away many of the obstructions, and the steamer, some weeks later, was able to move more easily. . . . Steaming past the river Ruo (an affluent of the Shire River) and through the Elephant Marsh, then frequented by large herds of elephants and buffalo, the country of the Makololo was reached. . . . On the arrival at his village of the *Lady Nyasa*, Chipatula, another important Makololo chief, sold us a quantity of ivory, which he kept stored on a small island. . . . Further on, Katunga was reached, the head of the lower river navigation. Here we established an important transport station of the Company, which existed until navigation was rendered impossible owing to the shallowness of the river.[21]

None of this part of Africa, in the 1870s and 1880s, was marked by political boundaries. To the north of Elephant Marsh, there was tribal warfare, and slave traders were still sending hundreds of slaves each year to the coast of East Africa. Meanwhile, in Britain, the work of David Livingstone and the missionaries who came after him, and the horror stories of the slave trade, were exerting a strong influence on public opinion. Likewise, there was the desire to prevent Portugal from extending its influence from the eastern coast of Africa across to the western side of the continent, which would have cut access between the British spheres of influence in southern Africa and eastern Africa. However, the Berlin Conference of 1885 established the right of Britain to administer the region as a Protectorate around the lake, and for British subjects to travel freely along the Zambesi and Shire rivers. Once British jurisdiction had been established, Britain, through its missionaries, traders and administrators, was able to bring peace to the country, abolish the slave trade, establish a government, and begin development of the region. British influence extended southwards along the Shire to about 17°S, and the first port of entry to the Protectorate was at the Ruo junction, where, during the coming years, the township of Chiromo was built.

Chiromo began to develop soon after Harry Johnston first arrived on the east coast of Africa in July 1989 as Consul for Mozambique.[22] At this time, relations between Portugal and Britain were strained and the Portuguese were sending expeditions into the country around the Shire as a means of establishing their influence and hampering settlement by the British. In August 1889, Johnston went up the Shire River for the first time. Close to the Ruo junction, he met Major Serpo Pinto, who was leading a Portuguese 'scientific' expedition in the lower Shire, and he then proceeded through Elephant Marsh to Katunga, where he met John Buchanan. The following months were tense while both Britain and Portugal were laying claim to the area around the Shire and Ruo rivers. In 1890, after considerable diplomatic activity in Britain and Portugal, the Portuguese accepted the British claim to the area north of the junction of the Shire and Ruo rivers and to the Shire Highlands.[23] Britain proclaimed the Protectorate of British Central Africa, an area of some 118,000 square miles. John Buchanan, as Consul, raised the British flag at Chiromo and burned the huts that the Portuguese had built during their occupation. Chiromo now became the entry point for travellers to the Protectorate and developed into a small town of considerable importance. It was a good location because a road was soon to be built leading directly from Chiromo to the Shire Highlands, and hence there was no longer any need for boats to travel further north through Elephant Marsh, often an impossibility when the water level was low.

Johnston returned to Chiromo in July 1891 and stayed there for several weeks, planning the basic essentials for the administration of the Protectorate, including such issues as communications, roads, treaties with chiefs, the role of the police and army for maintaining peace, and justice. Little has been written about Chiromo in the early days in spite of its importance for customs, communication and defence. It was a place that most people passed through on their way to the highlands and the lake. In his fascinating book *British Central Africa*,[24] Johnston described his impressions of the little town in an imaginary letter dated 12 June 1891:

Chiromo is an awfully pretty little place. The roads are broad and bordered with fine shady trees planted close together. Some of the buildings are quite smart, though of course at home we should think them small. Up to the present, the climate has been lovely and I have not had a touch of fever. It is quite cool at nights and one seldom gets mosquitoes, but I am told that in the rainy season they are an awful pest. In the middle of the day it is about as hot as a summer's day at home, but not too hot to walk about with or without an umbrella. This is the beginning of the cool season of the year.

While at Chiromo, Johnston spent time drawing an accurate and detailed map of the Shire River from Chiromo to Katunga, with details of villages, sandbanks and other geographical features.[25] In one corner of the map is Tshiromo, shown as a single road (some 50 yards in length) parallel to the Ruo. It is lined with houses, with a few little roads at right angles to the main road. There are also a few houses close to the northern bank of the Ruo. Immediately across the Ruo in Portuguese territory is a cluster of buildings, and there is another cluster, presumably an African village, on the west bank of the Shire opposite the Ruo junction. In its heyday, Chiromo must have been a delightful outpost of the Empire.

One of the most detailed and poetic of descriptions of river travel and the town of Chiromo is provided by R.C.F. Maugham [26], who joined Johnston's administration in Zomba in 1894. He left Chinde at the mouth of the Zambesi in the *John Bowie* in 'mid-winter'. Travel in the *John Bowie* was not pleasant as Maugham recalled:

> My cabin, as I have said, was small. It was populous with rats, immense cockroaches nearly three inches long, and with . . . other things. There was no mosquito net; and a small piece of muslin, which was all I could obtain in Chinde, was far too tiny to serve any useful purpose. To the miserable sleeplessness produced by swarms of mosquitoes was added, as night advanced, another variety of discomfort caused by the dew which filtered through cracks in the roof, and dropped icily cold upon one from above.[26]

Maugham was a very observant person, and his book *Africa as I Have Known It* is full of interesting comments on the landforms, plants and animals that he observed from the steamer. After stopping at Port Herald, where he met Mr Galt, the Collector, he was introduced to typical African hospitality: 'Arriving at his dwelling, my host lost no time in producing the indispensable "sundowner", and shortly afterwards a more than usually welcome bath, and a change into clean raiment, were the forerunners of a well-deserved dinner and a good night's rest'.[27] After his good night's rest, Maugham continued his journey in a 30-foot iron houseboat, 'ironically named the Rapid' in which a platform had been made in the stern covered with a thin grass thatched roof. A night stop was made, and Maugham spent an uncomfortable night on a straw-filled mattress under the thatched roof of the houseboat, where he was plagued by insect pests and the chill of night mists. Early in the morning, one of the African boatmen, Tanganyika by name, made coffee, and Maugham was able to take in the beauty of his surroundings:

It was a perfect morning. With increasing light, the tranquilly-flowing river, reflecting the greyish blue of the early sky, took on a wonderfully pearly, opaline tint, against which the floating islands of detached vegetation looked almost black, silhouetted as they were against the almost imperceptible current. Behind the dark outline of the mountains, gradually increasing in brilliancy, a pale saffron flush grew in intensity, until at length a blazing shaft of sunlight pierced the thin morning haze . . . Flights of lazy-flying white egrets winged their way overhead. A large grey heron swooped down and stationed himself on the edge of the bank so recently quitted by the noisy pugnacious band of hippos. Duck and geese, changing their feeding places, described moving lines on the brightening sky, and the swallows and swifts appeared once more in their apparently purposeless, ceaseless flight.

At about half past ten we came in sight of Chiromo, whereupon my boatmen, evidently for the purpose of announcing our arrival, struck up a more than usually resounding chorus, and bent to their paddles until the foam positively flew from under our bows. We passed the entrance to the Ruo river, and a few minutes later I was standing on the top of the bank and shaking hands with Mr G. A. Taylor, the acting Collector of Revenue and Chief of Customs. Together, and followed by a numerous and curious throng, we proceeded to the official residence, where I remained for a pleasant lunch.

I found that the chief topic of conversation was Captain Hunt's record bag of lions. Mr Taylor told me that, due to the immense numbers of the buffaloes and other game in the spacious, neighbouring Elephant Marsh, lions were most unpleasantly numerous – to such an extent, indeed, that by night they frequently entered Chiromo, and it was no uncommon thing to see their spoor, as also that of leopards, not only in the roads, but in the dust of the garden paths close to the houses.

After recording that a policeman had been killed by a lion in Chiromo within the last few days, and that chickens and goats were frequently killed by leopards, Maugham continued:

I found Chiromo a pleasantly situated and neatly laid-out settlement. Most of the more important houses, approached from the main road by well-kept garden paths, backed on to the Ruo River. The roads themselves were bordered by budding avenues of, I think, Acacias and Eucalyptus trees. Evidently, a great deal of hunting was indulged in, for, on the verandahs of the houses almost without exception, one noted the heads and horns of buffaloes, waterbuck, and many other varieties, while the atmosphere here and there was redolent of the various stages of preservation of skins, heads, and other trophies . . .

Another traveller was H. L. Duff,[28] who arrived in Chiromo in early 1898 and who, like Maugham, found it be a delightful place:

With the familiar Union Jack flying over its neat houses, and with its broad roads and its avenues of Persian lilac, Chiromo is a welcome sight after the desolate beauty of the Shire below that point and the eternal sand and pebble beaches of the Zambesi. It is, moreover, a great centre of hospitality, the British and Portuguese elements fraternising in the most cordial manner, and vying with each other in their kind attentions to passing travellers. Close to Chiromo lies the famous 'Elephant Marsh', a great natural resort of wild animals, and not long since a chosen haunt of the mighty beasts from which derives its name. Messrs H. A. Hillier and H. C. McDonald, the officials then in charge of British Chiromo, were both of them very keen and successful sportsmen, and some of their trophies were enough to fill the hearts of less fortunate hunters with mingled admiration and despair. . . . On the whole, the journey up the Zambesi and Shire rivers was very interesting to me as a new-comer, but at the same time I am bound to add that it was neither speedy nor a comfortable one. In the first place, the heat was intense and at times only just endurable, the thermometer often marking as much as 110 degrees in the shade. Mosquitoes again were a great source of annoyance.

One of the biggest problems for the early travellers and the transport of goods was the fluctuating level in the Shire River. At times it was so low that steamers and boats were stuck for days on end on the sandbanks; at other times, especially during the wet season, the flow was so fast that the surrounding land was flooded, rowing was almost impossible, and great rafts of floating vegetation blocked the channels. So it was not surprising that the possibility of a railway line along the river bank was discussed soon after the Protectorate was declared. In 1895, the Shire Highlands Railway Company was formed, and in 1904 the first section of the railway, on the west bank of the Shire river from Port Herald to Chiromo, was built.[29]

A second section, from Chiromo to Blantyre along the valley of the Ruo and over the escarpment, was completed in 1908. This section involved constructing a bridge over the Shire River – a major feat of engineering in those days – which was high enough to allow steamers to pass underneath and strong enough to withstand the force of water and floating islands of vegetation when the water level was high. A consequence for Chiromo was that the Customs moved south to Port Herald and travellers no longer had a real need to stop in Chiromo after their arduous river trip. When Mrs Colville visited Nyasaland in 1908, only six years before Rodney Wood first arrived in Chiromo, she had a much easier journey than did the earlier travellers:

We left the houseboat (at Port Herald) about 6.30 A.M. and walked to the railway station, being attacked by myriads of mosquitoes en route; these pests were far worse here than on the Zambesi. The train started at 7 A.M., following the course of the Shire as far as Chiromo, the headquarters of the district, a neat well laid-out station facing the elephant marsh, and consequently a veritable hotbed of mosquitoes. An excellent breakfast and luncheon were served on the train. The change in climate freshened us up, and altogether we made a very pleasant journey, greatly enjoying the beautiful scenery of the Shire Highlands through which we passed. After passing Chiromo, the line mounted gradually until we reached the watershed about 5 o'clock at an altitude of four thousand feet. . . . A short run of about half an hour down hill brought us to Blantyre about 5.30 P.M.[30]

It is difficult to visualise what Chiromo was like in the 1890s. One of the earliest photographs show a laterite road bordered by young trees and a huge baobab in the distance, but without any buildings.[31] Two other photographs show the first post office, a low, hut-like building with walls made of woven reeds, a roof of grass thatch, and wooden pillars supporting a thatched veranda. In one photograph, the first postmaster, Hugh Charlie Marshall, is sitting in a machila (a hammock supported by poles and carried by one or two people in front and behind), surrounded by various hunting trophies,[32] and in another, he is standing with nine smartly dressed African policemen wearing jackets, tasselled turbans and white gaiters, and carrying rifles.[33] At a later date, as Chiromo grew in importance, more substantial buildings were erected. The African Lakes Company built a very impressive two-storey wooden building, and large two-storey brick houses were built for the important government and commercial officials on the banks of the Ruo. There were also a number of shops run by Indians, and other buildings associated with the cotton trade and the railway. In 1906, the Church of St Paul was built by the UMCA to commemorate Bishop Mackenzie; it was a fine building built of local brick, with a tower and thin tall windows high up on the walls.[34] The church was consecrated on 11 February 1907.[35]

Boats and river transport played a major part in the life of Chiromo. Many different sorts of boats have passed up and down the river over the years and, early on, the settlement and development of Nyasaland was entirely dependent on the goods and materials which were transported up the river. The first steamer to pass the Ruo junction was the *Ma-Robert*, Livingstone's first boat during the 1858–1864 Zambesi expedition. She entered the Shire for the first time on New Year's Day 1859,[36] steamed 100 miles upriver during the first week of January, and reached the first of the cataracts north of Elephant Marsh on 9 January.[37] This extremely quick journey, at the beginning of the wet season, must have been possible because the river had risen sufficiently to cover the largest of the sandbanks yet was not flowing so quickly as to allow large moving islands of vegetation to block the river channels. For two years, the *Ma-Robert* steamed up and down the river carrying stores. However, she was a small uncomfortable ship, and was replaced by the larger *Pioneer* in 1860.[38] The *Pioneer*, likewise, made many trips up and down the Shire between 1860 and 1863. The next steamer to pass the Ruo junction was the *Ilala 1*, carrying Dr Laws and his party in 1875, as described above. In the following years, several steamers were assembled at the mouth of the Zambesi and regularly transported people and goods up and down the lower Shire throughout the later years of the nineteenth century. The best known of these steamers, whose names were recorded by early travellers, were the *Empress*, the *James Stevenson*, the *John Bowie*, the *Centipede*, the *Millipede*, and the (second) *Lady Nyasa*. In addition, there were two gunboats, HMS *Mosquito* and HMS *Herald*. They were all stern-wheelers and used wood for fuel; all called in regularly at Chiromo to discharge and take on passengers and goods and, then, when the river was high, continued upstream through Elephant Marsh to Chibisa or Katunga.

Perhaps the most amazing of the steamers to pass the Ruo junction were those which had been shipped out to Africa in thousands of pieces and were assembled above the falls at Matope. In addition to the *Ilala* (described above), at least nine others reached the lake this way by the turn of the century. The *Charles Jansen* and the *Chauncy Maples* belonged to the UMCA; the *Domira*, the *Livingstone* and the *Queen Victoria* were owned by the African Lakes Corporation; and HMS *Pioneer*, the *Adventure*, the *Dove*, and the *Guendolen* were the property of the government.[39] The effort and time expended to transport these vessels from Britain to Lake Nyasa is a good indication of the importance of steamers for many forms of work on the upper Shire and the lake. One example of the effort involved is provided by the history of the *Chauncy Maples*, a large ship measuring 127 feet long and 20 feet wide, and with a displacement of 250 tons. After being assembled in Glasgow,

> [she] was taken to pieces again and despatched to Africa in 3,500 pieces. The boiler, which could not be taken to pieces, was sent out whole. It was brought up the Zambesi on a barge from Chinde to Katunga. Thence it was successfully wheeled on a specially constructed carriage for 64 miles by 450 Angoni. . . [R]iver beds had to be crossed and hills climbed. It really was a very wonderful feat.[40]

Every single plate and bolt was numbered. Some 2,000 of the packages weighed about 50 pounds each (the weight of a single headload), others weighed between 50 pounds and about 500 pounds, and a few weighed even more. The boiler itself weighed nine and a half tons. There is a contemporary photograph of the boiler on its carriage being pulled along a dirt road by the long column of Angoni.[41] Rough measurements suggest that the carriage was about 20 feet long, and the column of Angoni who were pulling it stretched out in front for 50 to 60 yards. Building the road was an accomplishment in itself; it had to have a very slight gradient and the curves had to be wide because of the length of the whole column – some feat in such rough and hilly country.

By the time Wood lived near Chiromo, it was a flourishing town (see Chapter 5). All the large buildings were lived in and functional, the railway provided easy access to Blantyre and Port Herald, and river steamers were moored on the bank. Chiromo was especially busy during the First World War, when troops, vehicles, ammunition and many other goods passed through on their way towards German East Africa. But after the war, Chiromo gradually decayed. The Church of St Paul was moved stone by stone to Blantyre, a succession of floods swept through the town, and a particularly large flood in 1948[42] swept away the original railway bridge. Chiromo never recovered.

Cholo and the Shire Highlands

The Shire Highlands are situated above the eastern escarpment of the Rift Valley to the east of the Shire River. The escarpment, from just north of Zomba to Chiromo, is particularly well developed in this part of Malaŵi. It rises steeply from the Shire Valley (c.100–470 metres above sea level) to the highlands (c.1,000 to 1,300 metres above sea level). The Shire Highlands are a vast plain, originally covered with *Brachystegia* woodland savanna; the countryside is flat in some places (near Zomba, for example) and gently undulating in other places, with numerous small streams flowing through steep-sided valleys. The small town of Cholo (or Thyolo) is situated in the southern part of the Shire Highlands, 65 kilometres southeast of Blantyre and close to Cholo Mountain (1,462 metres). Because of its position above the escarpment and close to Mlanje Mountain, Cholo and the surrounding country has a cool, almost temperate climate and high annual rainfall (typically 1,000–1,600 millimetres per year although, close to Mlanje Mountain itself, rainfall may be a high as 2,000–3,000 millimetres per year). The soils are rich and good for growing a wide variety of crops, including maize, tobacco, coffee, tea, and many sorts of fruits and vegetables. In the early days, Cholo was rarely mentioned; it was not on the routes favoured by explorers or missionaries. As more farmer-settlers arrived in Nyasaland, and as the agricultural potential of the Cholo region was realised, coffee was planted extensively. In time, tea replaced coffee as the most widespread crop. The visitor travelling through the Cholo area today sees mile upon mile of neatly clipped tea bushes, and there are now only a few remnants of the original vegetation. Tea has become one of Malaŵi's most valuable crops.

Wood was attracted to the pleasant climate of Cholo after several years of living in the very hot climate of Chiromo; he made several collecting expeditions up the escarpment from Chiromo during the First World War, and then decided in 1919 to settle on a small estate among the Cholo hills (see Chapter 4).

CAPE MACLEAR

Cape Maclear is at the tip of a large peninsula of rocky hills which juts out into the southern end of Lake Nyasa. Just to the west of the peninsula is a sandy beach, some four kilometres long; it faces north-west and is protected at both ends by rocky outcrops. In the 1850s, it was a quiet, remote place where otters played among the rocks, shoals of multi-coloured fish swam in the sheltered waters, fish eagles sailed and cried overhead, and many baboons lived in the rocky woodlands on the surrounding hills. Livingstone and Kirk were the first Europeans to see this peninsula, from the deck of the *Ilala* on 5 September 1861, and Livingstone named it after Sir Thomas Maclear, the Astronomer Royal at Cape Town who taught him how to take geographical observations using a chronometer and a sextant.[43] Six years later (in September 1867), Lieutenant E.D. Young and Henry Faulkner spent two days on the beach while investigating reports that Livingstone had been murdered in a nearby village (a report which proved to be untrue). Faulkner recorded that he shot reedbuck and 'gazelles' (probably meaning impala), saw lots of guineafowl and monkeys, and found the spoor of buffaloes and elephants.

Eight years later, Cape Maclear was to change for ever when Young returned there as the leader of the expedition for the establishment of the first Livingstonia mission. The mission was composed of very practical people; it was their knowledge and skills which enabled them survive in such an isolated place so far from home, and with little contact with the outside world. There was Young himself (a naval man, skilled in all things nautical), Dr Robert Laws (a missionary doctor), George Johnston (a carpenter), John Macfadyan (an engineer and blacksmith), Allan Simpson (second engineer), Alexander Riddell (an agriculturalist) and William Baker (seaman, Royal Navy).[44] They decided to build the mission at Cape Maclear since it appeared to be a healthy place (although this proved not to be the case), it was near some of the routes taken by the slave traders, and it allowed easy access to the lake but was well sheltered from storms. The mission settled at the western end of the beach, close to the promontory which is now called Otter Point. During the next few years, eight substantial houses were built of brick, wood and thatch (one of them with two storeys); in addition, there were numerous storehouses, craft shops and schoolrooms, a large garden, fields for growing wheat and oats, a fort, a brick factory, and a drainage cutting which drained the marsh at the back of the mission. At the height of its development, in about 1880, there were 590 people living in and around the mission. But it was not a healthy spot; there were mosquitoes and tsetse flies, and the soil was poor. Many of the missionaries succumbed to fevers and other diseases, some had to be sent home, and several died.[45] In 1881, Dr Laws decided to move to another site, Bandawe, on the western shore of the lake much further north. The houses at Cape Maclear deteriorated quickly, and grass and trees reclaimed the land. There is nothing left of 'Old Livingstonia' now; only the large baobabs which still exist and were fully grown when the mission was established remember what it was like in those faraway days. Many years later, the little town of Monkey Bay was built to the east of Cape Maclear; Wood resided at Monkey Bay and Cape Maclear for 16 years in the 1940s and 1950s (see chapter 8).

LIVINGSTONIA

After leaving Cape Maclear in 1881, Dr Laws established the mission at Bandawe on the lakeshore halfway up the lake. But this site proved just as unhealthy as Cape Maclear, so in 1894, the mission moved to Kondowe, high in the hills above the lake.[46] At first, the new mission site was known as Kondowe, but it soon became the new 'Livingstonia'. At this time (six years before the Protectorate was declared), the northern part of the lake was still mostly unknown. A few miles to the north of Kondowe, Mlozi and his men were raiding villages and capturing slaves, many villages were abandoned and their inhabitants were hiding out in the bush.

The altitude of the Kondowe plateau is 1,300 metres. It was the perfect site for the mission: the air was cool and crisp, the soil was good, and there was plenty of water from streams and springs.[47] There was abundant clay for making bricks, and stone was quarried from a nearby rock face. Most importantly, the climate was healthy. Looking eastwards from the plateau, the hills tumble down steeply to the lake 850 metres below; to the west, the hills rise up to the open grasslands of the Nyika Plateau. Developing Livingstonia was akin to building a small town. During the following fifteen years or so, the mission built a manse, a church, houses, schoolrooms, workshops and a hospital. Gardens were established to provide vegetables, and the surrounding hills were planted with cedar, juniper and eucalyptus trees. A printing press was installed for the publication of Bibles, dictionaries and other literature, and a turbine generator powered by water provided electricity. Livingstonia still exists, and has been hugely successful by any standards because it was not just an evangelical mission; of equal importance was education, medicine and public health, and the training of artisans such as builders, blacksmiths, carpenters, gardeners, agriculturalists, roadmakers, tailors and many others. By 1914, some twenty years after its inception, the Livingstonia mission had expanded far and wide in northern Nyasaland; it is recorded that there were 49 European missionaries, 862 outstations and 57,000 children in mission schools, and nearly 27,000 patients had attended the dispensaries and hospitals.[48] Dr Laws remained at Livingstonia until 1927, although he and Mrs Laws went back to Scotland for two years between 1908 and 1910. The present-day Livingstonia is a tribute and memorial to the vision and determination of Dr Laws and his fellow missionaries.

Wood lived at Livingstonia, on and off, from 1938 to 1942 (see Chapter 7).

BLANTYRE

When the Free Church of Scotland was establishing its first mission at Livingstonia on the lakeshore, the Established Church of Scotland was looking for a site for its own mission. Rather than settle in a region close to the slave-trade routes, the mission decided that a location where the climate was healthier was more appropriate. Henry Henderson was entrusted to find a suitable place. In 1875, he travelled to Africa with Dr Laws and his party from the Free Church; he visited Livingstonia on the lakeshore, sailed up and down the lake, and then climbed up into the Shire Highlands.[49] David Livingstone himself had walked over these highlands between 1859 and 1861, and was impressed by the gently rolling hills, the mild tropical climate and the running streams. In early 1876, after walking over much of the highlands, Henderson camped beside the Mudi river, a lovely spot surrounded by seven hills. He decided that this was the perfect site for the mission because the area was well populated (and hence good for misssionary ambitions), the climate and soil were good, and there was abundant fresh water. In October 1876, the first missionaries – a doctor and five artisans – arrived. They established the Blantyre Mission (named after the Scottish town where Livingstone was born), laid out the square, built a wattle-and-daub church with a thatched-grass roof, a manse and bungalows, and made a road from the mission to Katunga on the Shire River. All this activity occurred some fifteen years before the foundation of the Protectorate, at a time when there was no law and order, and slave-raiding parties passed regularly through the highlands. Many remarkable people were associated with the mission in its early days; besides Henry Henderson, there was John Buchanan, who was appointed in 1876 as gardener-agriculturalist and who was responsible for introducing coffee into the country; Dr Clement Scott, who built the church; and Dr Alexander Hetherwick, who presided over the mission (and the development of the town of Blantyre) for nearly thirty years.[50] Traders and merchants were quick to settle in Blantyre; perhaps the two most influential and well remembered are John Moir and his younger brother, Frederick. They were born in Edinburgh, and after experience in East Africa, were appointed joint managers of the Livingstonia Central Africa Company (later the African Lakes Company).[51] They were a determined pair, almost ruthless in pursuit of their goals. John appears to have been a studious person; he wore spectacles – a great novelty in Africa at the time – and the way they sparkled in the sunlight earned him the nickname 'Mandala' by the local people. Frederick (as judged from his photograph taken in 1895) was typical of the explorer-hunter-empire builder type; he sported an immense dark beard, wore a topee and carried a gun. The Moir brothers established their headquarters about a mile from the Blantyre Mission in 1880; here they built a store surrounded by a stockage and protected by two circular defensive towers. In 1882, they built Mandala House for the resident manager and as a guesthouse for travellers.

In 1889, the Protectorate was proclaimed in Blantyre, and the Union Jack was raised on a flagpole near the mission. In the early days, Blantyre was formed of three separate parts: the mission, the government, and the traders. In the years before the First World War, Blantyre grew into a small town: it was laid out with streets, and many notable buildings were completed, including the European hospital (1897), the government 'boma' (1900) and the Queen Victoria Memorial Hall (1903).[52] One of the most significant events for the development of Blantyre was the completion of the Shire Highlands Railway, and the arrival of the first train from Chiromo on 31 March 1908. Contemporary pictures of Blantyre show that most buildings were single storey, constructed of locally made mud bricks and pale-coloured mortar, and with a corrugated iron roof and wide veranda. Roads were made of beaten earth, and transport was either by machila, rickshaw, bicycle or motorcycle. As the European population increased, and especially as the number of women rose, many social gatherings were organised including horticultural shows, croquet, tennis and cricket matches, horse racing and bicycle racing.[53] Of all the buildings in Blantyre which capture the feeling of the early days, the most impressive is the Church of St Michael and All Angels. The church was the inspiration of Dr Clement Scott, who was head of the mission, and his brother, Affleck Scott. There was no definite plan on paper (other than that there would be a chancel, two transepts, a dome at the east end, and two towers at the west end), and neither Dr Scott nor his brother had any experience of building. The church is constructed of mud bricks, locally made and fired. There was a lot of experimentation and the bricks for each section were first assembled without mortar, to check that the desired shape and effect was correct, before they were laid. One of the most remarkable features of the cathedral is that it is made from eighty-one different shapes of brick, a most unusual feature for a brick building, but it allowed for the intricacy of design typical of a stone-built cathedral. Each of the moulds for the bricks was carved personally by one of the brothers.[54] It seems that as the building progressed, the Scotts became more confident and experimental; by they time they reached the upper part of the building, they were designing ornate and complex roofs, towers and turrets. The church took about three years to build, and at the time of its consecration on 11 May 1891, it was the largest building in central Africa.

Many plants and trees were introduced to Blantyre (as well as to the rest of the country) from about 1876 onwards. The mission introduced many fruits, including apples, pears, oranges, lemons, plums, figs, pawpaws, Indian mangoes, strawberries and raspberries, all of which thrived well in the Blantyre climate. In contrast, gooseberries, blackcurrants and redcurrants were eaten by white ants and could not be established.[55] Many species of tree were introduced for shade and to provide wood. Paths and roads were lined with eucalyptus trees, and the fragrant perfume of eucalyptus oil filled the air.

By the time Wood first arrived in Nyasaland, Blantyre was a well-established thriving township, still very remote from other such centres, but large enough to generate its own activities and amusements. Wood never lived in Blantyre itself, but he often visited the town to buy provisions, collect money for paying staff at Magombwa, and stay with friends. Blantyre became the commercial centre of the Protectorate because of its location and its links with the outside world. A neighbouring settlement, Limbe, became the railway headquarters as well as a trading centre. During Wood's time in Nyasaland, Blantyre and Limbe merged into each other, increased in size and population, and gradually spread around the bases of the seven hills.

ZOMBA

Zomba is 69 kilometres north of Blantyre on the eastern flank of Zomba Mountain. Sir Harry Johnston chose it as the administrative capital of the Protectorate for two reasons: he thought the scenery was more beautiful than that at Blantyre, and he wanted to distance himself from the mission.[56] Zomba had already been chosen by Hawes in 1886 (when he was Consul) as a suitable place for his headquarters. Hawes bought a house and 100 acres of land from John Buchanan (who had settled in the district after leaving the Blantyre Mission). The house was called 'The Residency'and resembled a small medieval chateau; it was a large brick building with two storeys, a round turret capped with a conical roof at each end, and a grass thatched roof. (It still exists and has recently been modernised.) Behind the Residency, the steep slopes and rocky cliffs rise precipitously to Zomba Mountain. Lots of little streams, overhung by moist dense forest, flow down from the mountain, and the delightful sound of cascading water is never far away from anywhere in Zomba. In the early 1890s, Alexander Whyte, the Government Naturalist, planted a beautiful garden around the Residency and established the Zomba Botanical Gardens. Johnston kept leopards, servals, monkeys, a bushpig, crowned cranes and guineafowl in the grounds. Zomba was in 'the wilds' at this time: leopards and lions roamed at night, and many species of game lived on the Phalombe plains which stretched from Zomba eastwards to Mlanje Mountain.[57] One of the first government officials at Zomba was R. C. F. Maugham; many years later, he remembered Zomba with great affection:

> . . . of the many attractive spots that I have seen in Africa, Zomba is the loveliest of all. Situated in a wonderfully beautiful gorge some distance up the mountain, the Residency . . . looked out over extensive and tasteful gardens. The view, bounded in the extreme distance by the gigantic Mlanje mountains sixty miles away, roamed over a wide and fertile plain, wooded as far as the eye could [see]. Although between sixty and seventy miles away, on one of those phenomenally clear mornings of early summer, when the transparent atmosphere has been washed by the deluges of the early rains and the smoky haze of the grass fires has disappeared, the Mlanje range seemed so close that you would almost have sworn that the distance was not a quarter of what it was in reality.[58]

At the top of Zomba Mountain is an extensive plateau, shaped rather like a saucer, with higher peaks around the rim.[59] Here, 'the landscape is almost completely treeless and bears a striking resemblance to the moors of Derbyshire or the Yorkshire wolds – a resemblance heighted by its wide expanse of bracken'.[60] Because of its height, the mountain has high rainfall and is the source of many streams which flow off the mountain. In the cool season, it may be very cold on the top, with ground frosts at night. The views from the mountain are majestic. To the east is Mlanje Mountain; during the dry season, Mlanje is swathed in mist and smoke from burning grasslands, and cannot be seen from Zomba. At the beginning of the wet season, the mountain appears, as if by magic, like a pale, shadowy, ghost-like form and almost the same colour as the mists themselves. As the wet season progresses and the air becomes clearer, the outline of the mountain becomes harder, more detail is visible and the subtle brown and green colours of rocks and forests can be distinguished. Whatever the season, though, Mlanje always retains its air of mystery and magic – here one day, gone the next. From the heights of Zomba Mountain, the immense panorama of Mlanje and the changing pageant of clouds, mists, rainstorms and sun shafts above the surrounding plains provides one of the best sights in this part of Africa. To the west, the mountain plunges steeply down to the Upper Shire Valley, a wide plain of wooded savanna with the Shire River meandering like a silver thread in the far distance. On the other side of the valley, wooded hills merge into the blue haze of mist and cloud, and it is difficult to tell where the hills end and the sky begins.

Houses for government officials were built on the lower slopes of the mountain around the Residency. The terrain is so rugged that the paths and little roads had to zigzag along the contours and over little bridges. During the years before the First World War, Zomba developed as a small administrative centre with a hospital, military barracks, Government House, and a gymkhana club. Most of the people who lived in Zomba in the early days of the Protectorate were civil servants and military personnel. As a consequence, it did not develop the bustle and vibrancy of Blantyre; that, however, is one of its delights – it has remained a quiet, peaceful country town. Zomba was the perfect choice for the capital of Nyasaland: its good climate, large trees and bubbling streams, clear air and wonderful views made it one of the best places to live in the country, a reputation that it still retains to this day.[61] Wood lived in Zomba during some of his time as Chief Game Warden, and he visited it frequently at other times during his travels.

Chapter 3

EARLY NATURALISTS IN NYASALAND

The nineteenth century was an era of exploration and high adventure in Africa. Much of the geography of the 'Dark Continent' was unknown, except around the coastal regions. The maps of this era show the places where rivers enter the sea, but where these rivers originated in the hinterland and the route they travelled towards the sea was mostly guesswork and speculation. Likewise, the positions of lakes and mountains, from where rivers were assumed to originate, were equally speculative. It took many decades of exploration before the major geographical features of Africa were known, primarily due to the determination of great explorers such as John Speke, James Grant, Richard Burton, David Livingstone, Mungo Park, Hugh Clapperton and Richard Lander. Although these men were more interested in geography and mapping, they (and others) often made notes on the plants and animals that they encountered during their travels. Some of them collected specimens and sent them back to European museums. Over the years, a steady stream of pressed plants, pickled animals, skins, skulls and horns, and many other sorts of biological material reached the scientific institutions of Europe. As a result, the wonderful biological diversity of Africa started to become known to the scientific world.

Thus when Rodney Wood first went out to Nyasaland, he was not the first to collect animals in this part of Africa. Numerous species of flora and fauna had already found their way back to the British Museum of Natural History and to the herbarium of the Royal Botanic Gardens at Kew. These tended to be the most obvious and widespread species, and those that caught the attention of travellers because of their size, colour or unusual appearance.

The first specimens from the region that became Nyasaland were collected during the Livingstone expedition of 1858 to 1863. David Livingstone and some of his companions kept detailed diaries in which they recorded their day-to-day activities and thoughts; mostly these concerned the level of the water in the river, the difficulties of navigation, the weather, the endless squabbles between the members of the expedition, their frequent illnesses and diseases, and observations about local people and customs. Livingstone's journal, as might be expected, also included much on the horrors of slavery and the problems of converting Africans to Christianity. Sometimes the diaries, especially those of Dr John Kirk and James Stewart, include comments on the plants, mammals, birds and insects that caught their attention for one reason or another, as well as notes on their collections. Out of all the members of the expedition, it was Kirk, Stewart, Charles Livingstone (David's brother) and Dr Charles Meller who made the greatest contributions to the biology of this part of Africa in these early years.

John Kirk was a medical doctor by training, but he was a man of many talents. Besides reading medicine at Edinburgh University, he attended botany classes, learned some geology, became proficient at shooting, and learned the new art of photography. After graduation, he visited Turkey, the Levant, Egypt and Spain, where he collected plants and took photographs. So by the time he was appointed as medical doctor and economic botanist to Livingstone's Zambezi expedition, when only 26 years of age, he was already a well-travelled and accomplished young man.[1] The expedition originally planned to focus on the Zambezi, because of Livingstone's determination to use this river as a highway into central Africa. However, this was not to be; the extensive Kebrabasa Rapids upstream from the junction of the Zambezi with the Shire made navigation impossible, and carrying boats overland was out of the question. So, as an alternative, the expedition turned its attention to the Shire River.[2]

Kirk's collections and observations on the Zambezi, the Shire and the surrounding highlands covered the best part of five years. During this time, he travelled up the Shire River on six separate occasions. He also travelled further up country, climbed Mount Dzomba (Zomba Plateau) from the Shire Valley, visited Magomero where the expedition established its first missionary settlement, and he was with Livingstone when they first sighted Lake Nyasa. And, on one occasion, he steamed up the lake almost to the northern end.

Kirk recorded many anecdotal observations about the animals and plants which he saw around him, and he collected many specimens (which are still carefully curated in the British Museum of Natural History and at the Royal Botanic Gardens at Kew). In early 1859, Kirk and his companions steamed up the Shire River and he recorded that 'the Shire flows for 100 miles nearly north in a plain of about 20 miles wide. There is a district in the middle which is marshy and cut up into islands overrun by elephants; but the greater part is fine land for the growth of cotton, sugar-cane and rice'.[3] In the northern part of the marsh, there was a fine stand of palm trees, and Kirk recorded that 'we saw a herd of cow elephants which had just been eating the fruit of Palmyra. They suck it just like men, rejecting that hard nut.' On a later visit, he noted that:

> the elephants have retreated to the marshes at the foot of the hills where we see immense herds of them standing secure against attack. Where the mudbanks have been exposed, there are large flocks of waterfowl. On shore we found guineafowl and hornbill abundant. The hornbill had been eating beetles along with fruits . . .[4]

On 26 March 1859, Kirk and Livingstone shot an elephant in the marsh, in what seems a rather foolhardy manner, using a bullet full of poison. They followed the elephant amongst the long grasses, paddling their little boat in the shallow water. The elephant was a young female and when they eventually killed her, they took measurements and started to cut her up for meat with which to feed the Makololo tribesmen who were assisting the expedition. Kirk discovered that the elephant was pregnant, and wrote in his diary: 'The uterus I secured. The placenta consists of a host of rounded bodies scattered over the interior. The cord divided into three portions. The foetus was well grown with hair, no signs of teeth . . . We took on board a fearful amount of meat which the Makolo declare they will eat, also the head and the ear.' He also recorded that 'the foetus [could not be] adequately preserved because the following night a heavy shower came on and flooded me in my cabin, coming through the roof. It also went into my cask of pickled elephant and diluted [the preservative] . . .' All he was able to keep was the bones of the head and the placenta. Only a very dedicated collector would go to such lengths to preserve something that he thought would be of great scientific interest. After leaving the marsh, Kirk observed 'a great many guinea fowl . . . a few francolin and great flocks of bee-eaters' near Chibisa's village.

During his travels, Kirk collected plants and made stuffed specimens of the birds he shot. His diary is full of comments about these activities and the difficulties of preserving specimens. For example, on 19 August 1859, he wrote:

> Such a nuisance cockroaches are. They eat everything. My specimens are going under them. Corrosive sublimate however is a tolerable protection although it will eat the animal preparation, although the skin be washed with it, if there is any meat left under.[5]

And on 24 August:

> While we lay there [at Chibisa's], I had the opportunity of stuffing a few birds (Botany is dull, this is the dry time) . . .

Kirk was also good at observing and recording small details about his specimens, suggesting that he was a very good naturalist, interested in all aspects of the world around him. On 9 July 1861, he recorded the following:

> on stuffing the hornbills, the gizzards were found full of figs, and a variety of fruits, with the berry of a loganaceous tree which is used as a fish poison.[6]

And on 21 September of the same year:

> A new species of duck was shot here, smaller than the sort further South. [On this day, too, Kirk obtained a] goat swallow [in which] the feathers in each [wing] flow like a pennant behind. They are only to be seen about this season and for the next few months when it seems they lose their feathers.[7]

His observations were perfectly correct. Nowadays the goat swallow is known as the Pennant-winged Nightjar; during the breeding season, the male develops an extra long feather on each wing, bare of vanes for all its length except for the tip. When flying, the two pennants trail behind the wing, with the appearance of being unattached to the wing in front.

His observations were not confined to large species, as the following passages show. On his way to climb Dzomba (Mount Zomba), he 'observed a large grey fly depositing its young. These are perfect grubs, and at once move off, increasing visibly in size in the course of one minute. The fly is ovoviviparous. It had grey stripes on the back . . .' Kirk had evidently seen a tsetse fly giving birth to its grubs, which later bury themselves and turn into pupae. And on one of his visits to 'Lake Nyassa, a splendid lake [with] cranes [and] reeds on the bank', he observed:

> . . . swarms of small insects. They are carried about over the lake by the winds, forming clouds of deep brown colour and taking on many shapes as a cloud of smoke would do. When they alight, the people gather them and eat them. They are a sort of midge and very small. On shore, they are to be seen on the fir trees.[8]

These 'flies' (locally known as 'nkhungu') are a species of *Chaoborus* midge; their larvae develop in the mud at the bottom of the lake, and when development is complete the adults emerge synchronously, forming huge swarms.

Kirk had much difficulty keeping his collections in good order. Besides the cockroaches mentioned above, his camp was robbed one night: 'My bundle of papers was opened, the strings taken off, and the specimens pulled about, then heaped in any way and left. The fish skins were all thrown together. Botanical specimens were smashed by this rough treatment. The birds are gone.'[9] He must have been heartbroken by these losses; as any collector knows, specimens which are lost or damaged can never be replaced. Kirk himself rightly commented: 'People at home can have no idea of the thousand obstacles to making a collection and getting it home safely.' He could also have added that many people would have had no idea of all the equipment that he took to Africa with him to make his collections: guns, shot, cartridges, preserving solutions and powders, cotton wool (for stuffing the specimens), notebooks, insecticides, storage boxes, mounting and drying papers for plants, pins, labels, dissecting instruments and many other necessary items. This was not unusual – even collectors of today have to take similar amounts of equipment and encounter similar obstacles!

However, many specimens did eventually arrive in England, where they were identified and accessed into the collections of the British Museum of Natural History and Royal Botanic Gardens. Various mammals, including genets, palm civets, mice, fruit bats and insectivorous bats, as well as some fish and amphibians, 48 specimens in all, were accessed on 9 January 1863.[10] How many bird specimens from the expedition finally found their way to the Museum in uncertain, but it seems that a total of 276 specimens arrived, over several years, at the British Museum of Natural History.[11] The number that were actually collected is also uncertain because many were destroyed by insects or damp, or were stolen, while still in Africa, and some may have been lost on their way to England. Kirk also amassed a huge collection of plant specimens, most (or all) of which were sent to the Royal Botanic Gardens.

Kirk had the ability to encourage other members of the expedition to participate in his activities. Charles Livingstone, much maligned by his dominant brother, was also a competent naturalist and helped Kirk to collect and prepare specimens. Many of the specimens attributed to Kirk were probably collected and prepared by Charles Livingstone.[12] With hindsight, it is perhaps fairer to note that the two men jointly made a significant collection under extremely difficult conditions rather than to apportion too much credit to either one or the other.

Collecting some specimens also had the advantage of providing food for members of the expedition, many of whom were good shots. For example, on 22 February 1863, Kirk wrote: ' Dr Meller and I . . . sometimes we get a goose or a pair of pretty little ducks, the best game in the country, or it may be a black heron. Knob-nosed ducks are to be had occasionally. All these serve as good food and make the table at the dinner hour more pleasant than when pressed meat or salt ordinary rations are the sole condiments. Besides, a few birds pass the day and their skins fill the collections.'[13] On Christmas Day 1861, Kirk joyfully recorded: 'We had a jolly Christmas dinner both fore and aft. Seven men forward could not finish a black goose while we went in for small geese and duck.' No doubt the skins of all these birds became specimens, but in addition Kirk had the foresight to record what the birds had been eating: 'The food of the knob-nosed goose killed yesterday had been *nyamba* or small beans. The small ducks had been eating the fruit and roots of the *Nymphaea*.'[14] At other times, meat from elephants, hippopotamuses, various antelopes, many species of birds, turtles and fish gave variety to their diet.

In addition to these members of the expedition, there are two others who merit attention: Dr Charles Meller and Dr John Dickinson. Both were medical doctors; they did not write their own diaries (or least their diaries are no longer available), and so they are much less well known than John Kirk and Charles Livingstone. Meller joined the expedition in 1861 as ship's doctor to the *Pioneer*, although he also spent some time at Magomero, where the Universities Mission set up its first mission on the Shire Highlands, not far from the present town of Zomba.[15] He was also a naturalist and scientist of some ability.[16] He collected plants and animals, and made the first survey of malaria in central Africa which was subsequently published in the *British Medical Journal* in 1862 and in the *Lancet* in 1864.[17] At that time, the transmission of malaria by mosquitoes was not known, but Meller appreciated the concept of malarial immunity (especially among Africans), and he described the different symptoms of the disease. He even recommended that people should avoid being bitten by mosquitoes and that a full dose of quinine should be taken at the onset of the disease. He was a good and caring doctor, but the overwhelming effects of disease in central Africa and the deaths of members of the expedition (notably Bishop Mackenzie, Richard Thornton and John Dickinson) had a very sobering effect on him. He evidently was very concerned about the health of members of the expedition – every one of them was seriously ill at one time or another – and believed it was prudent that anyone with health problems should leave the country.[18] In this respect he was at odds with many of the others, especially David Livingstone, who were determined to battle on regardless. Meller himself went to South Africa to recuperate from illness in April 1862, but returned again in October 1862. In spite of the criticisms directed at him by David Livingstone and John Kirk (who appreciated Meller's talents as a botanist), Meller's contribution is significant because of the large collections of plants that he made with Kirk in the Shire Highlands.[19] He also collected a rather special mongoose while he was at Magomero.[20] Charles Meller was an accomplished artist (a talent that the better-known members of the expedition did not comment upon), and he painted a series of delightful watercolours of expedition life, each full of detail and action.[21]

Dr John Dickinson was the first resident medical doctor of the Universities Mission, and he was stationed at Magomero.[22] He arrived in November 1861, together with Bishop Mackenzie and the Reverend Burrup. The catastrophes at Magomero and the early deaths of Mackenzie and Burrup have already been recounted (see Chapter 2), and are not part of this particular story, but by April 1862, the mission had moved to Chibisa's on the Shire River, on Dickenson's recommendation. Dickinson was said to be 'delicate',[23] but even so, he looked after the medical needs of the ten Europeans and 105 Africans at the mission. By mid-1862, however, it was clear that he could not survive the climate of the Shire Valley and the inadequate diet, but he refused to leave until another doctor could replace him. He is quoted as saying: 'If God spare me I shall be thankful; if it be His duty to take me, let me die in the performance of my duty.'[24] In spite of his weakness due to fever and other diseases, and his duties as medical officer, Dickinson collected birds (118 specimens, including many predatory species), spiders, and diatoms,[25] and spent many hours watching birds on the Shire River. He died on 17 March 1863, before a new doctor arrived to replace him, and he was buried close to the river near the mission.[26]

Harry Johnston is better known in the annals of African history as the first Commissioner of British Central Africa and (later) of the Protectorate of Nyasaland. Like Kirk, he was a man of many talents: on the memorial in the little church at Poling in Sussex, where he is buried, he is described as an 'administrator, soldier, explorer, naturalist, author and painter'.[27] He was a very determined person, with exceptional abilities. His contributions to the natural history of Africa are immense. They are especially impressive when it is realised that much of his time was occupied with ending the slave trade, settling tribal conflicts, making treaties with local chiefs and setting up the infrastructure of government. However, in some people's eyes, he also had his faults. R. C. F. Maugham, for example, referred to Johnston as lacking presence and dignity, and as being effeminate and somewhat eccentric. Nevertheless, his legacy of specimens, books and drawings are an enduring testament to his energy and enthusiasm.

Johnston's first experiences of Africa were during 1882 and 1883.[28] During these years, he travelled in West Africa and Angola, collecting specimens and making sketches and paintings. He also became interested in the politics of Africa. In late 1883, he was appointed as leader of the Kilimanjaro Scientific Expedition, sponsored by the British Association and the Royal Society. His collections during the expedition were presented to the British Museum of Natural History and are the subject of several papers published in scientific journals. By the time he first travelled to the Shire River and Lake Nyasa, he was already an accomplished African collector and traveller.

Johnston steamed up the Zambezi and Shire rivers for the first time in July 1889, when he was the British Consul for Mozambique.[29] He explored the Shire, the lake and the surrounding highlands for about a year before returning to England. For the next few months, while in England, he was involved in delicate diplomatic negotiations with Portugal regarding the fate of the country surrounding the lake and the Shire River. In the end, Portugal withdrew its claim to the country, and Johnston was appointed as the first Commissioner to the Protectorate of British Central Africa. He in turn appointed many dedicated and adventurous men to the new administration, including Alexander Whyte, who had the delightful title of 'Naturalist to the Protectorate and Head of the Scientific Department'.[30] This was a brilliant appointment. The two men, in spite of their very different ages and backgrounds, must have got on well together. Whyte was about 60 years of age when he went to British Central Africa; Johnston was only 30! From his photograph, Whyte appears to be a formidable old man, white-haired with a moustache and goatee beard, and large dark eyebrows; in contrast Johnston looks young and impressionable. Whyte had been a tea planter in Ceylon for much of his life and was a horticulturist by training. In Africa, he collected in most parts of the Protectorate – an amazing accomplishment for a man of his age. Whyte stayed in British Central Africa until 1898, two years after Johnston left, and then went to Uganda to join him there. Between the two of them, they made huge collections of many animals and plants. Their collections included mammals, birds, reptiles, amphibians, land shells, butterflies, flowering plants, ferns, liverworts, fungi and lichens.[31] Besides collecting, Whyte designed and planted the garden at the Residency in Zomba and helped to look after Johnston's little 'zoo'.[32]

Johnston's contribution was not confined to collecting alone. In his monumental book *British Central Africa*, published in 1897,[33] he devotes 179 pages (or about one third of the book) to botany and zoology. He describes many interesting indigenous plants, where they occur, and other characteristics of their biology and usefulness. For example: 'Fibre is also obtained from the aloes, baobab and the arboreal *Hibiscus*; the extraordinary *Kigelia* tree (whose seed pods are sometimes nearly as thick as a man's thigh and like a huge pendant sausage in shape) contains in its seed pods a fibrous material like the Egyptian Lufah which can be used for rubbing the skin after a bath, and might be utilised for many other purposes. The natives take the seed of these *Kigelia* pods and roast and eat them in times of scarcity. A species of hemp, probably introduced, grows wild all over British Central Africa. It is smoked by the natives . . . and it might also be got to yield a fibre, and some of the palms would do the same.' Johnston also comments on species that provide drugs and good timber. The section on botany includes two appendices; the first (by Harry Kambwiri, a scholar at one of the Church of Scotland missions) lists all the local names of trees of British Central Africa which are of value to humans; and the second (by I. H. Burkill of the Royal Botanic Gardens) lists all the plants known to occur in British Central Africa at the time.

The chapters on zoology in British Central Africa are full of similarly interesting descriptions and details, and contain many of Johnston's paintings and drawings of mammals, birds, reptiles and fish. Most of the descriptions are of the larger and more spectacular vertebrates and insects, and for many species Johnston includes details on distribution and ecology. Such information is easily found in many books and guides now, but in Johnston's day, such details were mostly unknown. His comments on buffaloes are typical of the breadth of his interest and knowledge. After recording the range of the buffalo throughout Africa and commenting that the forest buffalo of West Africa is a 'degenerate' type, he continues:

Buffaloes are very abundant all over British Central Africa, but of course are retiring from the vicinity of European settlement. They are also frequenters of the plain rather than the mountains though they will ascend high plateaux in the dry season for the sake of the green herbage. The favourite places of their resort are wide marshy districts like the Elephant Marsh near Chiromo, where even after the most wanton and indiscriminate slaughter at the hands of Europeans, they exist in large numbers – thousands, it is said. Like the Indian buffalo, they are fond of wallowing in mud and water, though perhaps not as aquatic in their habits as the last named animal. They are dangerous beasts to tackle under certain conditions though less dangerous than the elephant and lion. It is seldom that they will take aggressive action against the sportsman when not wounded.[34]

Johnston also provides checklists of the mammals, birds, reptiles, amphibians, fish, land and freshwater molluscs, and several orders of insects then known from British Central Africa. No other book on Africa published at this time (or for many years afterwards) provides such detailed information about animals and plants.

One of the most valuable aspects of Johnston's descriptions is that he recorded the abundance and distribution of mammals in the Protectorate at the end of the nineteenth century which can now be compared with similar information at the end of the twentieth century. Such comparisons are sobering, and clearly show the adverse effects of hunting, environmental change and increases in human populations during the last 100 years. This topic is discussed elsewhere (Chapter 6), but two examples will serve to show the value of Johnston's records. Johnston wrote that 'the zebra of British Central Africa is a singularly beautiful beast [and] is still extremely common almost all over the Protectorate, and measures have now been taken to preserve it from undue diminution at the hands of sportsmen and natives.' One hundred years later, only a few small populations survive in four national parks and game reserves (Nyika Plateau, Vwasa Marsh, Kasungu and Nkhotakota). The Nyasaland Wildebeest, a particularly beautiful form of wildebeest, was never a common species in British Central Africa, and Johnston recorded that it was found only 'in the vicinity of Lake Chilwa and the Elephant Marsh'. The species has not occurred in these localities for many years because of excessive hunting and it is now locally extinct, although small populations may still be present in northern Mozambique.

Johnston travelled extensively. Most of his travels were connected with his duties as Commissioner, but he always seemed to have time to collect specimens and to paint and draw. A constant stream of specimens was sent back to England, and as with those sent back by Kirk, they had a long and hazardous journey. Specimens were packed in boxes and carried by porters to the lower Shire River at Katunga, where they were placed on one of the paddle steamers and taken down the Shire and Zambezi rivers to the coast at Chinde. Here they were stored until a passenger or cargo boat arrived; and then the sea journey to England took about six weeks. The specimens could easily have suffered from damage or loss, but amazingly most of them, if not all, survived the journey unharmed.

The mammalian specimens collected by Johnston and Whyte were identified almost immediately on arrival in England and they formed the subject of a series of scientific papers in the *Proceedings of the Zoological Society of London* by Dr Oldfield Thomas, one of the foremost mammalogists at the time.[35] From the papers, and by looking at the specimens themselves, it is possible to retrace Whyte's travels during his years in British Central Africa. As might be expected, many specimens come from the area around Zomba where Johnston and Whyte were based. But in addition, Whyte collected at Tschiromo (Chiromo), Mulanje, Fort Lister, Mpimbi, Malosa, Chiradzulu and Fort Johnston during the period 1891 to 1895. In 1896, he made an extensive tour to the north of the Protectorate and collected specimens at Monkey Bay, Fort Hill, Karonga, Masuka (Misuku Hills), Livingstonia and the Nyika Plateau. Until a few months before Whyte's travels in the north, the infamous Mlozi and his agents had continued their nefarious trade in slaves in the lawless region around Karonga. In December 1895, Johnston, accompanied by a large army of Sikhs and African troops, defeated Mlozi in a battle at Karonga; Mlozi was captured and hanged. Peace and stability thus came to the north, allowing Whyte to make the first collections from this biologically very interesting part of the country.

The outstanding features of Johnston and Whyte's zoological specimens are the care and attention that was paid to the preparation of the specimens, and the information contained on the labels. The mammals and birds were superbly stuffed so they resembled the correct shape and proportions of the animals when they were alive; and the labels recorded, correctly, the exact locality where each specimen was found, the date, the appropriate measurements, and sometimes other information of interest. It is fascinating to look at these specimens now and to speculate on the conditions and hardships that must have been endured in order to obtain them. No doubt, local Africans were employed to collect some specimens and perhaps to help with their preparation and preservation. Whyte must have been a very determined collector, as shown by an amusing incident recorded by Dr James Henderson, a missionary at Livingstonia.[36] When Whyte stayed at the mission, he visited the nearby Manchewa Falls, which have extremely steep, rocky slopes covered by wet vegetation. In his quest for specimens, he tied a length of calico around his waist that was then held by two men while he was lowered down the slopes. It was quite a feat for a man of 62 years of age!

The collections were presented to the British Museum of Natural History by Johnston in his capacity as head of the administration. The labels on most of the specimens have a printed heading with 'Nyasaland H.H. Johnston, C.B.' at the top, although after Johnston received his knighthood, it was changed to 'Nyasaland Sir Harry H. Johnston, K.C.B.' At the base of some of the labels is printed 'A. Whyte, Collector' or 'Alexander Whyte, F.Z.S. Naturalist'. The labels suggest a slight hint of the pompous streak in Sir Harry's character (which others have commented upon) and his sometimes inadequate acknowledgement of Whyte's contributions. However, Oldfield Thomas, in his papers published in the *Proceedings*, paid full credit to Whyte's abilities; in the first paper, of 1892, he wrote:

> . . . a fine series of mammals sent home . . . by Mr H.H. Johnston, C.B., F.Z.S., Consul-General for Mozambique and H.M. Commissioner for Nyassaland [sic], under whose auspices they were collected by Mr Alexander Whyte, F.Z.S., a trained naturalist and collector on Mr Johnston's staff, who is engaged in investigating the fauna and flora of Nyassaland. It is impossible to speak too highly of the scientific energy and public spirit of Mr Johnston in furthering our knowledge of the natural productions of the region which he is called upon to govern . . . Mr Whyte, the actual collector, also deserves special mention for the energy with which he has carried out the work entrusted to him and for the care and attention which he has devoted to the preservation of the specimens.[37]

And in the fourth paper in the series, published in 1896,[38] Thomas wrote: 'As usual, the majority of Sir Harry Johnston's specimens have been obtained by that indefatigable naturalist Mr Alexander Whyte.' Finally, in the fifth paper, one year later, he gave his final valediction:

Now that Mr Whyte has retired from his labours in the tropics, it is only fitting that in this, the last paper that will appear on his mammals, special reference should be made to the great value of the services he has rendered to zoology in general, and to our knowledge of mammals in particular, and to the way in which, during the last six years, he has utilised the opportunities given to him by the generosity and public spirit of Sir Harry Johnston.[39]

The contributions of both Johnston and Whyte were published in many zoological and botanical papers and publicly acknowledged by the Zoological Society of London. In 1897, Whyte was awarded the silver medal of the Zoological Society for 'valuable services to zoological science by his researches in British Central Africa'.[40] He was also elected a Fellow of the Linnean Society. Johnston was awarded the silver medal 'for investigations in British Central Africa' in 1894, and later (in 1902) he was honoured again by the award of the gold medal of the Society 'in connection with his discovery of the Okapi'.[41]

Many other keen naturalists, better known for their other contributions during the early days of British Central Africa, collected animals and plants during their spare time. Although their collections were not so noteworthy, all were of value because they provided new knowledge on distribution and biology, and sometimes included species that were new to science.[42] These naturalists were a varied lot; they included Sir Alfred Sharpe and Sir William Manning, who succeeded Sir Harry Johnston as Commissioner or Governor; John Buchanan, a planter on the Shire Highlands who introduced coffee into the Protectorate and built the Residency at Zomba where Sir Harry lived; Captain Bertram Sclater of the Royal Engineers (the son of Dr Philip Sclater, Secretary of the Zoological Society of London) who built the first roads in the Protectorate; Dr Percy Rendall, a medical officer in the administration;

Captain E.L. Rhoades, a naval officer on the lake during the Battle of Karonga in 1895 (when Mlozi was defeated), and who was better known for winning the first naval battle of the First World War by destroying the only German naval vessel on the lake; and Dr S.A. Neave, an entomologist who collected widely in central Africa. This list is by no means exhaustive, but it serves to show that during the years succeeding Livingstone's expedition and before Wood's arrival in Nyasaland, many species of animals and plants were recorded from the country. There were, of course, many gaps in knowledge; most collecting had been around the stations and missions and in localities of special interest, such as the mountains and the lake. In contrast to the value of the collections made by these naturalists, many animals mostly mammals and birds were shot by hunters for food and sport, often in great numbers. In terms of science, these 'specimens' were completely wasted; a few were of sufficient size to be recorded in hunting books, but that is all.

Once the specimens reached Britain, they were accessed into the collections and identified. In many instances, scientific papers were written describing each collection, or noteworthy specimens from a collection, with an analysis of the value and interest of the specimens to science. Species new to science were described in detail and given a scientific name. There is a very well-defined set of 'rules' for biological classification and for naming a new species. The following brief description of how this works may be helpful to the reader, and is necessary to fully understand Rodney Wood's contribution to biology. Any species of animal or plant is classified according to its class, order, family, genus and species, a hierarchical arrangement referred to as the Linnean System of Classification. For example, the Bushbuck, a beautiful small antelope that lives in woodland thickets in Nyasaland, belongs to the class Mammalia, order Artiodactyla, family Bovidae, genus *Tragalephus*, and species *spekei*. The scientific name of the bushbuck is simply the genus and species name: *Tragelaphus spekei* (with a capital 'T' and a small 's') and is written in italics. Any specimen considered to be a new species has to be named in this way. The scientific name is unique to that species; a genus may contain more than one species and all species in a genus share a set of common characteristics implying that they are related, in an evolutionary sense, to each other. For example, there are seven species in the antelope genus *Tragelaphus*, including the Bushbuck, the Nyala and the Kudu; all of them have spiral horns and other features of the skull that are unique to the genus. A species name may be associated with more than one genus name, but in this case, no relationship is implied. The name *capensis* (meaning 'from the Cape', although in the nineteenth century anywhere in South Africa was considered as the Cape) is the specific name for the Honey Badger (*Mellivora capensis*), the Cape Hyrax (*Procavia capensis*) and the Clawless Otter (*Aonyx capensis*). Within the animal and plant realms, there are dozens of organisms with the specific name *capensis*, but each is associated with a different genus name.

A characteristic of the naming process was to use names, especially for the species name, which described either the animal itself or the locality where it was found, or to honour the collector or some other personage. Using mammals and birds of Nyasaland as examples, locality names include *shirensis* (after the Shire Valley), *zombae* (after Zomba), *nyikae* (after the Nyika Plateau), *nyassae* (after Lake Nyasa), and *angoniensis* (after Angoniland). Descriptive names include *ruber* (reddish colouration), *albiventris* (white ventral region) and *tricolor* (three colours). Perhaps the most interesting, in a human context, are names referring to individuals who are associated with Nyasaland. The name *johnstoni*, for example, has been used for six species or subspecies of mammals in Nyasaland: a wildebeest, two species of shrews, a dormouse, a wild pig, and a hyrax.[43] In addition, there are species of birds, fish and amphibians, as well as invertebrates, named after Sir Harry. Likewise for Alexander Whyte: a monkey, a rodent-mole, a small climbing mouse, and a hare,[44] as well as some birds, fish, insects and molluscs.[45] Whyte's name is also associated with the best known of all the endemic trees of Malaŵi, the Mulanje cedar, which has the delightful scientific name of *Widdringtonia whytei*. Other interesting names include *Rhynchogale melleri* (a mongoose named after Charles Meller), *Raphicerus sharpei* (a small antelope named after Sir Alfred Sharpe), *Heterohyrax manningi* (a hyrax named after Sir William Manning) and *Falco dickinsoni* (a falcon named after Dr John Dickinson). The binomial scientific name, as we now understand it, is a wonderful invention, especially when it describes the magnificence or uniqueness of the species or the locality where the species lives, or when it provides an historical link with the individual who was responsible for finding the first specimen known to science.

However, names may not always remain in use; taxonomists (biologists who specialise in the classification and naming of animals and plants) may discover, through examination of additional specimens or reappraisal of the evidence, that a species or subspecies can no longer be considered to be valid and that the 'species' is in fact just a variety of another species described earlier elsewhere. By the rules of nomenclature, the earliest name is used and any later name becomes a synonym and is no longer used. This has happened to many of the names which were used to describe animals and plants during the nineteenth century and the early years of the twentieth century. However, many species still bear the names of localities in Nyasaland and commemorate the early naturalists.

The individuals described in this chapter were the first in a continuous stream of naturalists and biologists who have contributed to our knowledge of the flora and fauna of Nyasaland. By the time Rodney Wood arrived in Africa, the larger and more obvious species of animals and plants had been given scientific names, and some information was known about them. A few books had been published that could be used for identification, such as Boulenger's four-volume *The Fishes of Africa*, Lyell and Stigand's *Central African Game and its Spoor* and Sclater's *Mammals of South Africa* (all of which were owned by Wood), and descriptions had been published in a few scientific papers (most of which were unavailable in Africa). The early naturalists, by their collections and writings, provided an important basis for Wood's studies. However, there were still many animals and plants about which nothing was known, and more or less anything that could be collected was likely to be new and interesting – the perfect situation for a keen young naturalist such as Rodney Wood.

Chapter 4

FIRST YEARS IN AFRICA: 1909–1921

Rodney Wood travelled to Africa for the first time in 1909 to work on a farm in Southern Rhodesia. There is no record of why he made the decision to forego a future in the wine trade or of whether his father approved of this decision. It seems highly likely that the business world of London and the day-to-day chores of trading in wine were not to his liking and that he longed for a life of freedom and adventure so he could indulge in his passion for natural history. The only record of Wood's first years in Africa is the short comment: '[In] 1909–10 [I worked] on the farm of James Watson, Esq., Salisbury, S. Rhodesia'.[1] James Watson was the owner of an estate, Kilmuir, in the Enterprise region, about twenty miles north-east of Salisbury (now Harare). This region was settled by pioneers who arrived in 1890; many of them became farmers and were allocated farms of about 3,000 acres. James Watson was a leading light in local society; he was a founding member of the Enterprise Farmers' Association and its Secretary for many years. He was a proud Scot, dogmatic in his views, and he did not have a high opinion of the English![2] A local publication recorded the frugal side of his character: 'James Watson certainly made money. He made it by not spending it. To part with sixpence was grievous to him. After his death, Joe Read, who rented Kilmuir, found his diaries and was amused to read one entry: "Went into town today, met ——, had lunch at the Club didn't cost me a bean!"'[3]

Wood spent only a year or so at Kilmuir; he then went to Canada for nine months, 'recovering from malaria and tick fever' and gaining agricultural experience.[4] However, his severe illnesses and long convalescence did not diminish his enthusiasm for Africa, for in 1911 he returned to take up an appointment with the British Cotton Growing Association (BCGA) in Northern Rhodesia and in Nyasaland and, later, with the British South Africa Company in Northern Rhodesia.[5] The BCGA had its headquarters at Mazabuka, a small town about fifty miles southwest of Lusaka. Cotton had been grown successfully in these regions for many decades – the Livingstone expedition noted that there were little plantations of cotton along the Shire Valley in the 1860s – and hence the colonial government encouraged planters and local farmers to grow more cotton and to plant other crops of economic importance.

During his three years with the BCGA, Wood had the opportunity to travel and he started to collect specimens and take an interest in local history and culture. In 1912, he was posted from Mazabuka to Port Herald on the Lower Shire and from there he went to Vua, a small village on the lakeshore south of Karonga in northern Nyasaland, to begin a small plantation producing cotton seed.[6] This short visit was Wood's first experience of Nyasaland, but he saw enough of the country and its people during his stay to fall under that magical spell which affected so many Europeans and caused them to remain there for all, or most, of their lives. His first collections, in 1913 and 1914, were made around Chilanga and Mazabuka (both in Northern Rhodesia, now Zambia). They included fruit flies – very small flies which congregate around rotten fruit[7], and chironomids – small, delicate flies whose larvae develop in mud at the bottom of ponds, lakes and rivers.[8] Most people would hardly notice these lowly insects. Many of Wood's specimens turned out to be new species: one of the fruit flies was named *Chryosoma woodi*,[9] the first of many species that were named in his honour. He also began to collect small mammals, and in November 1913 he obtained a specimen of an insectivorous bat at Chilanga that he sent to the British Museum of Natural History. It proved to be a new species of slit-faced bat and was named *Nycteris woodi*,[10] or Wood's Slit-faced Bat. Wood was also interested in aquatic biology, as are many budding naturalists, so while he was stationed at Port Herald in 1913, he visited Chiromo, where he collected a few fish from the Shire River. There was nothing new or special about these fish, all being known already from further downstream, but they provided evidence that they did occur as far upstream as Chiromo. In these early days of collecting, any record from a new locality was of interest. Wood, like any young collector, must have been very pleased to learn of the value and importance of his collections, and no doubt was encouraged to make larger collections at every available opportunity.

While in the Port Herald and Chiromo regions, he started to collect samples of local craft and to learn the language. Many years later, in 1958, he recorded his impressions:

> As I was much interested in acquiring a collection of carved wooden household goods, still used and made by the Mang'anja of the Port Herald district, I spent much time wandering among villages, getting to know the people and trying to improve my grasp of the language, of which I already had a working knowledge acquired during the past year in Northern Rhodesia.
>
> Among other things I was particularly interested in their bows and arrow, and archery still much practised by them. (I was to see a lioness killed by massed arrow fire later). Very soon the older men and I had some competition shoots together, using their old original form of bow and reed-shafted arrows, and I learned how deadly they could be up to some sixty yards distance.[11]

Wood stayed with the BCGA until 1914, when he resigned because he 'saw no promising future there'.[12] He then joined the staff of a commercial cotton estate near Chiromo, where he stayed until 1920. The estate was owned by John H. Lloyd of Edgbaston Grove, Birmingham, although it was managed by Lloyd's son. Wood recorded that his house was

> . . . right on the bank of the Ruo river, five miles north of Chiromo . . . [and] our plantation here is made in the heart of dense forest . . . bats are exceedingly numerous here . . . [It is] quite flat, but the hills start one mile further north . . . [the Rift Valley escarpment leading to the Shire Highlands]. Here the plain is all forest, but northwest of Chiromo is the fairly open Elephant Marsh . . . There never being any really cold weather here, insects are very numerous throughout the year, except for two or three months, but even then they are not quite absent.[13]

Referring to the Shire River, he recorded that '. . . [a] characteristic of this area for fully eighty miles along both banks of the river's course is the preponderance of *Hyphaene*, and in certain places *Borassus* palms throughout the open forest and woodland, bordering the river and its great swamps, or scattered in clumps throughout the latter, extending

71

back several miles from the river'.[14] During the years when Wood was working for the estate, the locality reference on many of his specimens was 'Chikonje'. This was the name of a railway siding where cotton bales from the estate were loaded on to the railway.[15] Chikonje, which is not shown on any modern map, was some six miles east of Chiromo at a little distance from the northern bank of the Ruo River, near the present village of Makanga. The nearest town to the estate was Chiromo, one of the first of the European settlements in Nyasaland and an important place in the history of this part of Africa (see Chapter 2).

Cotton was an important commercial crop during the early days of Nyasaland. It was introduced into the country as a 'European crop' in 1900 and for a number of years headed the list of exports in terms of value. However, following this, crop yields declined and many estate owners replaced cotton with tobacco. In 1916–17, there were 27,300 acres of cotton and 9,300 acres of tobacco on commercial estates; by 1930–31, however, cotton was planted on only 760 acres, while tobacco had increased to about 19,000 acres.[16] During this time, many local African farmers had also started to grow cotton, and this partly made up for the decline in production on the estates. One of the major causes for the decline was the detrimental effect of insect pests, primarily from the caterpillars of the cotton bollworm moth which destroy the flowers. In 1916, losses were very high, and the Government Entomologist recorded that Wood participated in an experiment which removed bollworms by hand-picking them from some 45 acres of cotton on the estate at Chikonje.[17] Little information is available on Wood's employment on the cotton estate; however, it is hardly surprising that he decided, at the end of the war, to leave, and to grow tobacco and tea on his own estate instead.

By the time that Wood settled into life in Chiromo, it was a small, thriving township with a number of Europeans who worked in the cotton industry, on the railways and river steamers, and for the government. Chiromo was hemmed in by water and swamps: the Shire River on the west, the Ruo River on the south, and Elephant Marsh on the north. The single main road was lined with trees, and by a number of Indian shops, as well as by the African Lakes Corporation Office and the Church of St Paul. Most of the large European houses were built on the bank of the Ruo River, or close to it, where the large trees and proximity to water ameliorated the very hot climate. To the west of the town was the railway bridge over the Shire River (which was wide enough only for the tracks, but was probably also used by motorised vehicles – mostly motorcycles – and walkers), and to the east was a dirt road which climbed the escarpment to Cholo, Luchenza, Limbe and Blantyre. In the years immediately preceding the First World War, and during the war itself, it would have been relatively easy to travel up and down the river by boat and train, and to reach the highlands by foot, bicycle, motorcycle or train. Chiromo was not a healthy spot, and many of the early travellers, administrators and sailors died of malaria and were buried in the small cemetery on the west bank of the Shire.

Life in Nyasaland at the time when Wood arrived must have consisted of a mixture of great challenges, adventure and hard work. Those who coped and survived had to be fit and self-reliant, and able to endure isolation and hardship. Most of the services and entertainments available in Britain at the time did not exist in Nyasaland, and people had to make their own amusements. It took several weeks to reach Nyasaland from Britain. First there was the long sea journey either through the Suez Canal or around the Cape to Chinde, a small British enclave in the Portuguese territory at the mouth of the Zambezi. Large sandbars prevented ships from docking at Chinde, so passengers on their way to Nyasaland were placed in a large wickerwork basket and lowered into a waiting tug that took them ashore. At Chinde, passengers stayed in a boarding house until a little sternwheel paddle steamer took them up the Zambezi River to the mouth of the Shire. Frank Winspear, who travelled to Nyasaland in the early years of the twentieth century, recalled:

these steamers only drew about 18 inches and so could travel in very shallow water. The boiler, engines, saloon and cabins were all on the main deck and below that there was no place for cargo and baggage, so our heavy luggage was carried on barges lashed to the side of the ship. As the name implies, her propelling paddle was at the stern of the vessel and enabled it to pass through narrow channels, though of course in order to enable this to be done the accompanying barges had to be fastened and towed behind. The fuel was wood and stops had to made at intervals to take on firewood. The cabins were all very small and we all found it difficult to get any sleep in the hot nights. . . . Although the Zambezi and Shire are broad rivers, navigation is by no means easy. There was only one channel deep enough for the stern wheeler and its barges and this was not easy to trace. A man was posted in the bow of the boat to watch the course and to advise the captain when we seemed to heading for a sandbank. It was not unusual to strike a sandbank and then the barges had to be unfastened from the side and towed off the bank.[18]

At Chiromo, 'then a busy place', Winspear found there were no porters to take him and his party to Blantyre, so they continued up the Shire through Elephant Marsh in much the same way as had been done in the days of Livingstone some fifty years earlier. It took three days to reach Katunga from Chiromo, and then an evening and most of a night travelling in a machila to reach Blantyre. However, with the opening of the Shire and Highlands Railway from Port Herald to Chiromo in 1904, and from Chiromo to Blantyre in 1908,[19] the old style of travel was gone for ever. The train ran twice each week, leaving Port Herald at 7.00 a.m. and arriving in Blantyre at 5.00 p.m. the same day. The rapid development of Nyasaland is reflected in the number of ocean boats which stopped at Chinde on their way to and from England. The Union Castle Line operated several steamships, such as the *Dunevegan Castle* and the *Llandovery Castle*, which took passengers from southern Africa each month directly to England via the Suez Canal and Gibraltar. An alternative route was by steamer from Chinde to Durban, and then by steamship to England. Likewise, the German East Africa Line had northbound mail boats which left every two weeks, calling at Suez, Naples, Marseilles, Tangiers, Lisbon and Southampton, and southbound mail boats which called at Cape Town and Southampton.[20] A journey from London to Blantyre on the German East Africa line cost £60.2s.8d (first class) and £43.12s.8d (second class). There were also other shipping lines with less expensive rates: for example, Rennie's Aberdeen Line cost only £45.10s.0d (first class) or £36.0s.0d (second class) from London to Blantyre. Passengers who wanted to travel to Karonga (in the north of Nyasaland) could, for example, travel by Rennie's Aberdeen Line from London to Chinde, and then by river steamer to Port Herald, rail to Blantyre, machila to Fort Johnston, and lake steamer to Karonga for about £63 (first class) or £53 (second class).[21]

Life in Nyasaland during Wood's early days was to some extent determined by communications, activities generated by the local community, and what could be grown or imported. Chiromo was definitely an outstation, even though many people passed through it on their way upcountry or to the coast. In the early days, walking was the main way to get from place to place, although the machila was a popular alternative means of transport. As roads improved, rickshaws (two-wheeled carts pulled by a single person, very similar in design to the rickshaws of the Far East) and bicycles provided greater speed. The *Nyasaland Times* of 1914 contains advertisements for Triumph, BSA and Humber bicycles for about £10 (or slightly more for a three-speed model). By 1914, the first motorcycles were used for longer distances; they could be purchased or hired, and there was even a 'cycle car' motor service (taking two passengers) between Blantyre and Fort Johnston.

There were many imported goods in the shops, and local markets provided plenty of fresh vegetables and fruit. The advertisements in the *Nyasaland Times* and other contemporary publications give some idea of what was available: fresh meat, butter (2/- per pound), fresh milk (2d per pint), 'standard flour' (3/6d per 14lb bag), fine flour (17/6d per 56lb bag), sugar (3½d per lb), cooking oil (1/3d per bottle), Ceylon tea (2/- per lb), Nyasaland tea from Thornwood or Lauderdale Estates (1/- per lb), pillowcases (1/3d each), sheets (4/- each), cigarettes (5/- to 7/- per 100), cigars (4/- per box of 25), fizzy lemonade (5/- per 2 dozen bottles), Walker's whisky (72/- per case of 12 bottles), business envelopes (7/6d per 100), ebony rulers (2/- each), and office pins (1/- per box). Stores such as the Central African Trading Company, the London and Blantyre Supply Company, and the Kabula Store sold all manner of agricultural machinery and tools, cement, corrugated iron, nails, pressure lamps and furniture. With a little ingenuity and effort, a house could be made quite comfortable and homely. Visitors could stay at the Mandala Hotel in Blantyre or at Hodges Hotel in Zomba for 10/6d per day or £3.3s.0d per week. In Blantyre, there was the Sports Club, which offered entertainment and a golf course.[22]

Life in Chiromo must have been rather simple for Wood, because there were only a limited number of things that he could do, and items that he could buy. His house on the banks of the Ruo[23] would have been made with locally made mud bricks, and roofed with iron sheets or grass thatch. It was small, and judging by how he lived in later years, it would have been simply furnished with tables, chairs, bookcases and a bed. Like all houses of the period, it would have had a veranda and an outhouse with a tin bath. The kitchen, with a small wood-fired stove, was probably close by, and water would have been stored in buckets and 'debbie tins' (old four-gallon paraffin and petrol tins). At night, the house would have been lit by several paraffin pressure lamps. Outside there would have been a simple garden with a bird table (he always liked to attract birds to his houses) and perhaps an area for vegetables in the wet season, and all around was natural savanna forest. Nearby was the Ruo River, and a large colony of Crimson Bee-eaters that nested in the banks of the river.

Life in the Lower Shire Valley is dictated by the climate. After rising at dawn, Wood probably spent all morning supervising work on the cotton estate. In the early afternoon, when the estate workers went home, he would have returned to his house and spent the hottest hours of the day indoors. In the cool of the evening, he might have watched birds along the river and collected specimens. After dark, by the light of the hissing pressure lamps – there was no electricity and no radio in those days – he would have written notes and letters, catalogued specimens, and read books until it was time to go to bed. Bats (some of which ended up as specimens!) flew into the house. All Europeans hired local staff to look after their houses, to buy meat, fruit and vegetables from the local market, and cook meals. Every so often, Wood must have made expeditions into Chiromo, five miles distant from his house, to purchase supplies from the local Indian shops and the African Lakes Corporation store, and to post and collect his mail at the post office. No doubt he met with other Europeans and listened to the latest news, looked at the river steamers moored on the bank of the Shire, and purchased a copy of the *Nyasaland Times*. During those years, Wood learned the local language, Chinyanja, so he was able to converse fluently with the estate workers and with the local people who helped him collect specimens.

During the war and until 1921, Wood collected many specimens (see Chapter 11). His energy and determination were amazing, especially in view of the intense heat of the dry season and the hot and humid climate of the wet season. Under such conditions, most people would just sit and do very little! Although his duties on the cotton estate kept him very busy,[24] he still seemed to find the time to collect and curate his specimens. Looking at the dates and localities where Wood collected his specimens, it is evident that he travelled endlessly back and forth throughout southern Nyasaland. Collecting specimens is not easy; it requires knowledge of where and when a potential specimen may be found, and once the specimen is acquired it must be preserved and labelled correctly. In the tropics (more so than in cooler climates), specimens decay and go mouldy if preserved incorrectly, they may be eaten by one of many types of insect pests, or they may just fall apart because of the heat and humidity. Labelling each specimen is an essential but laborious task, but any specimen without details of its locality, date of collection, and collector's name is of no value at all. Additional information, such as habitat, time of collection and other biological information (including abundance of the species, time of day when collected, and whether the specimen was found alone or in association with others), makes a specimen even more valuable. Wood must have spent many hours and days just writing labels for his specimens, and it is because he did this that his specimens are of such value to biologists. Finally, specimens had to be sent home to a museum for safe keeping, involving a long and hazardous journey by land and sea to England.

During his first years in Southern Rhodesia and Nyasaland, Wood collected specimens of many groups of animals. He compiled a fine collection of insects, including many biting flies and ticks (which he probably caught when they tried to bite him). All these insects were sent to the Imperial Bureau of Entomology in London, and were subsequently deposited in the British Museum of Natural History. Wood presented about 4,500 insects from Northern Rhodesia and Nyasaland during the period 1913 to 1921.[25] Some insects, such as biting flies, blackflies, bees, beetles, mosquitoes, fleas and fruit flies, were well represented in the collection; others, such as bugs, weevils and dragonflies were poorly represented. Wood was never fond of spiders, so this group, too, is not well represented. Of all the insects, butterflies and moths were Wood's favourites. Why he found butterflies so appealing is not known – maybe it was their beauty, or their wonderful diversity of life forms, or all the subtle differences between and within species. Maybe his interest in butterflies was kindled at Ardvreck or Harrow, and in any case butterfly collecting was a popular hobby in England when Wood was a boy. He used many methods to catch butterflies for his collection.[26] Most specimens were caught with an insect net – many people who knew Wood in Africa commented that he always carried one when outside, just in case a butterfly that he wanted happened to be passing by. He also placed bananas on the ground and in trees; as bananas rot, they produce an odour that is very attractive to many species (including the large and difficult to catch *Charaxes*) which come to feed on the fruits. Other methods included placing urine on the ground (both the moisture and the odour are attractants) and rearing adult insects from caterpillars. His first-known butterfly specimen (still in existence) was collected in Ruo Gorge (close to Chiromo) on 15 March 1915, and he continued to observe and collect butterflies for most of his life.[27] Wood was well known in Nyasaland for his intense interest in butterflies; even in the 1990s, thirty years after his death, an old-timer remembered him by his nickname of 'Butterfly Brain'!

In addition to butterflies, Wood's other passion was birds. He was an ardent birdwatcher and collector, and in his day he was probably the most knowledgeable ornithologist in Nyasaland. During the First World War, he started to make a collection of bird specimens, recording the details of each one in a large foolscap-sized book.[28] Across the top of the double page, he wrote the column headings in his careful, distinctive handwriting: collection number, specific name, English name, sex, date, altitude, locality, colour of eye, colour of legs and feet, length (tip of beak to end of retrices, in millimetres), stomach contents, field notes, disposal, and index. By present-day standards, this is not particularly remarkable; but for a collector in central Africa in the early years of the twentieth century, such attention to detail is astonishing. His first specimen was a Great Tree Duck, collected on 27 January 1917 on the Ruo River at Chiromo. The majority of his 459 specimens were collected between 1917 and 1919; thereafter, he collected only occasionally and spasmodically – a few in 1920, between 1926 and 1928, and in 1940 (see Chapter 11). Wood also collected the eggs and nests of many of the species of birds in southern Nyasaland, and recorded the number of eggs/clutch and the characteristics of the nest.

Wood made a wonderful collection of small mammals in the Ruo district (primarily near Chiromo and Chikonje) and at Cholo during this period. Although some of the species were known and had already been described from other parts of Africa, Wood's collection was the first from this part of Africa. Between 1913 and 1921, he sent 258 specimens – 122 rodents, 109 bats, 11 elephant shrews, 9 shrews, 4 mongooses, 1 pangolin and 2 bushbabies – to the British Museum of Natural History.[29] He sent his specimens in small consignments over several years, partly to minimise the chances of loss in transit, but also so that he could receive provisional identifications and learn which species were of special interest. In August 1922, Philip Kershaw published his important paper on Wood's collection. In the introduction, Kershaw wrote: 'this interesting collection of beautifully prepared specimens is the result of the labours of some years, and adds very considerably to our knowledge of the distribution of the small mammals of the district'.[30]

One of the many skills that Wood learned at Ardvreck and Harrow was the importance of good observation and good records. When he sent his small mammals to England, he also sent detailed notes about many of the species. Kershaw included extracts from these notes in his 1922 paper; in fact, most of the paper is comprised of Wood's notes. On *Hipposideros commersoni* (a large insectivorous bat, similar in size to a fruit bat), Wood wrote:

> When a large species of wild fig, known locally as 'mtundu'-tree, ripens its fruits all along the stem of its branches, these bats come around in hundreds, like swarms of fruit-bats, land on the tree and seize the fruits, fragments of which are scattered by them all around, and are often carried to other trees nearby and pieces dropped there. The locals state that they eat the fruit, and call them by the same name as the true fruit-bats, i.e. 'mleme'. I wrote this to Mr Oldfield Thomas who replied that no *Hipposideros* was a fruit eater. On examination of the figs, I found that practically every fruit was attacked by a large weevil, the larvae of which were inside the fruit. It is therefore probable that it is on these weevil larvae that the bat is really feeding and that they only seize the fruit to tear it apart to get at the larvae.[31]

On another species of bat, the Yellow-bellied Vesper Bat (*Scotophilus nigrita*), he wrote:

> I have only found the genus *Scotophilus* in hollow or large holes in *Hyphaene* palms. The forest of the low country around the Shire River (Ruo and West Shire districts), and also that round Lake Nyasa and the Upper Shire River, is full of these *Hyphaene* palms. In such places *Scotophilus* is very common, and as many as twelve or twenty are sometimes got out of one hollow palm, which they inhabit together with all species of 'free-tailed'

bats (*Chaerephon, Tadarida, Mops*, etc.). I have never found them in any other species of tree, but they probably inhabit hollow *Borassus* palms as well where these are found. They are often noticeable at dusk hawking cotton bollworm moths and other insects over cotton-fields cleared in this type of forest, where the hollow dead palms have been left standing, and in this way they must do a lot of good.[32]

Such observations were rarely made by early collectors, but are of great value to biologists.

When Wood first went to Africa, there were very few publications that enabled him to identify any of the animals he obtained for his collections, especially the smaller species, and there was no resident in Nyasaland who could give advice. However, books on large game and hunting were available, and he had his own copies of Stigand and Lyell's *Central African Game and its Spoor* and the seventh edition of Rowland Ward's *Records of Big Game*.[33] Obtaining the equipment and preserving materials in Nyasaland that he needed to make specimens was almost impossible. Because of these difficulties, he established good working relationships with two experts at the British Museum of Natural History – Dr Oldfield Thomas of the Mammal Section[34] and Dr W. L. Sclater of the Bird Room.[35] Wood corresponded regularly with both these men, especially with Thomas, who strongly encouraged him to collect, helped him with equipment, and identified the specimens that were sent to the museum. Wood's letters to Thomas were handwritten in ink and were written in numbered duplicate books. The letters give some idea of the difficulties of collecting. On 2 June 1916, for example, Wood wrote:

> I have now managed to train a local man to prepare the skins in accordance with your instructions sent me, so am now collecting fast. Am very glad have managed to train a man to do this fairly well, I think, as I have not time myself for anything but the measuring and labelling. Am keeping a record of all specimens caught and measured, so that if any labels get lost I still have the particulars.[36]

Wood was anxious that all his specimens were identified; but he also wanted a small reference collection for his own use in Nyasaland, so he arranged for a tin box to be kept at the museum for specimens that were to be returned to him. And he added: 'In the event of death, my collection would go to the museum.'

Specimens of small mammals may be preserved either as stuffed dry skins (which resemble the natural shape of the live animal) or in some sort of preserving fluid. Preservation of specimens in Africa was a real problem, and it is highly likely that many of Wood's dry skins were destroyed by mould and insects. To begin with he used wood ash rubbed on the inner surface of the skin, as well as allowing it to dry in the sun (not a good idea because the skin becomes inflexible and brittle). Later, after Thomas had sent arsenical soap and labels, Wood and his local assistant were able to make excellent well-preserved specimens. The second method of preservation, in a fluid, is advantageous for some species, especially bats, because it preserves the shape of the animal as well as the internal organs. But alcohol and formalin, two commonly used fluids, are expensive and bulky. In one of his letters, Wood commented that whisky was the only spirit obtainable in quantity, in addition to petrol and paraffin, and wondered whether it would be suitable for preserving animals.[37] His correspondence with Dr Sclater followed similar lines[38] – requests for arsenical soap and boric acid powder, identification of specimens, the return of specimens not required by the museum, and acknowledgement of advice.

Wood was very good at just sitting quietly and watching the wildlife that passed by, as shown by his observations on elephant shrews:

Both *Rhynchocyon* and *Petrodromus* have the curious habit of striking the ground sharply with the tail, so as to produce a rapping sound. In the dense thickets it can be heard all day if one listens carefully for it. I have frequently sat and watched them doing this only a few yards from me, and often several will be doing it at the same time near each other. Sometimes one individual will stop and rap every few feet. They often appear to listen after it, but not always. . . . They stop at each puff of wind, as if suspicious, and hold the head up in the air, with the curious mobile tip of the nose moving about testing the wind, and reminding one forcibly of an elephant's trunk doing the same.[39]

And of the diminutive Pygmy Mouse, he recorded that it was:

very common throughout the highlands, where it is found in holes in the ground and among refuse, particularly in native gardens and maize fields and fallen grass. It is also very common in the open type of 'msuku' (*Uapaca kirki*) forest found all over the highlands of Nyasaland. It is said by the natives often to close the mouth of its burrow with small stones, and it stores grain in chambers in the burrow. Its native name is 'Pido'.[40]

Wood's methods of collecting were very simple because he had very little equipment. For rodents, he used old-fashioned mousetraps (exactly the same ones as can be purchased now), as well as capturing (by hand) any animals that he found in holes, burrows and nests. Bats were captured with an insect net, or by hand while they were roosting during the daytime, and occasionally by shooting with very small shotgun pellets. Birds were obtained mainly by shooting with a shotgun. Some specimens were purchased from local people, some of whom had a good knowledge of their fauna. However, Wood was often frustrated by the lack of interest shown by local people, even though he provided a monetary incentive for any specimen that was brought to him. Nevertheless, he employed and trained a 'collector' who prepared many of the skins very admirably. Insects were captured by hand or with an insect net, and by placing baits on the ground. All specimens, when dry and labelled, were stored in tin boxes until shipped to England.[41]

The First World War changed Nyasaland from being a colonial back-water to being an essential supply line for the war effort in German East Africa. From early August 1914 until November 1918, nearly a quarter of a million soldiers and others in Nyasaland were engaged in the war effort. Most of the effort was directed at crushing the German forces in German East Africa and preventing them from invading and occupying Nyasaland and the lake. Soldiers of many nationalities, together with stores, travelled up the Shire River and on the railway to the highlands and the lakeshore. Base camps and field hospitals were established at Fort Johnston (where the Shire River leaves the lake at its southern end), at Bandawe and Mbamba Bay (on the eastern lakeshore) and at Zomba, the main headquarters of the King's African Rifles.[42] The Germans never managed to establish themselves in Nyasaland, primarily because of two skirmishes. The first was on 13 August 1914, only a few days after war was declared. The only British naval boat on the lake, HMS

Guendolen, commanded by Captain Rhoades, located the German gunboat, the *Hermann von Wissmann*, in dry dock at Sphinxhaven on the German East African side of the lake. Before the Germans were aware that anything was amiss, the *Guendolen* scored a direct hit on the *Wissmann*. The captain of the *Wissmann*, Captain Berndt, who was a good friend of Rhoades, jumped into a dinghy and rowed furiously out to the *Guendolen* to find out was happening. His famous words – 'Gott for damn Rhoades: vos you dronk?' – have not been forgotten. Apparently Berndt had not been told that Britain was at war with Germany, so it was no wonder that he was furious and perplexed when his gunboat was shelled. The second incident occurred on 8 and 9 September 1914, when a German expeditionary force (comprised of German officers and African troops) entered Nyasaland from the north and marched towards Karonga. Unbeknown to the Germans, a strong force of the King's African Rifles had arrived in Karonga at the end of August and had fortified the town. The outcome was that the German troops were defeated and fled back across the Songwe river to German East Africa.[43] For the remainder of the war, within East Africa, Nyasaland's role was to provide troops for the East African campaign, and to participate in harassing General von Lettow-Vorbeck all over German East Africa. The people of Nyasaland were well informed about the progress of the war in German East Africa and overseas. The *Nyasaland Times*, now published weekly rather than twice each week, contained telegraphic news from Reuters, commentaries and editorials on the war, and news from the Nyasaland Field Force. There were many reports to show that life went on as normally as possible: the Blantyre Tennis Club still held tennis and cricket tournaments; the Blantyre Kinema advertised films (2/- for officers and civilians, 1/- for soldiers); Sunday services were held in the cathedral in Blantyre and at Zomba Church; the train and boat services still operated (although it was necessary now to have a passport before booking a trip to England); and the advertisements showed that it was still possible to buy unlikely items such as shaving brushes, wheelbarrows and brandy.[44]

The exact date of Wood's arrival in Nyasaland in uncertain, but he may have arrived from Northern Rhodesia shortly after the *Guendolen* incident. The earliest definitive record is for the collection of a fruit fly (which turned out to be a new species) at Chiromo on 23 September

1914.[45] He was soon involved in the war effort, because he enlisted in the Port Herald section of the Nyasaland Volunteer Reserve (NVR) on 8 February 1915.[46] The NVR was a volunteer movement that was started in 1900 with the object of providing a reserve force of trained marksmen for service in case of emergency.[47] Members of the NVR were issued with government weapons (usually .303 rifles) and 300 rounds of ammunition per year for practice. Each volunteer had to provide himself with a khaki uniform, but after 1917 uniforms were provided by the government including yellow shoulder straps embossed with the letters NVR in black.[48] The NVR, although composed of non-military volunteers, worked closely with the King's African Rifles and even participated in some actions against the enemy (for example, at Karonga). It also helped to quell the Chilembwe rising in January 1915, when the Reverend John Chilembwe, a priest of the Providence Industrial Mission, and his associates, killed three British planters near Magomero. The uprising was short-lived and the reprisals were swift and widespread, but it caused considerable panic amongst the European community at the time. By February, the uprising and its aftermath were over [49] and Nyasaland concentrated on its war effort.

Wood was a member of the NVR for most of the war. Exactly what he did was uncertain, but he recorded that he was a private and involved in the lines of communication.[50] Undoubtedly his knowledge of bushcraft and the local language would have been of great value. He also recorded, in his letters to Thomas, that in 1917 'war work and supervision of [the] plantation' prevented him from doing any trapping and that 'military work now takes me up to the Cholo range of this Ruo district so I hope to get many more species [of small mammals] there'. And in January 1918, he commented that the 'war still drags on out here and it is just possible I may get a commission in the Cape Corps, but nothing certain yet – depends on casualties'.[51] Wood's deployment to Cholo and the uplands near Mlanje was no doubt due to the 'cat and mouse' tactics of von Lettow-Vorbeck, who had moved south from German East Africa into Portuguese East Africa adjacent to the border with southern Nyasaland. Events changed for the better towards the end of 1918, and on 29 August he wrote to Thomas: 'Thank God for good news from Europe. This fool campaign out here goes on as usual. I am definitely discharged now as totally medically unfit, so have more time now, though very busy on the plantation.' His discharge certificate commended him for 'very good work on Lines of Communication'.[52] The armistice in Europe was signed on 11 November 1918, although von Lettow-Vorbeck did not surrender until 14 November 1918 at Kasama in northern Rhodesia.[53] For his services to the war effort in Nyasaland, Wood was awarded the Africa General Service Medal (as were all members of the NVR).[54]

Nyasalanders made a significant contribution to the war effort; they formed the ranks of the local regiments of the Kings African Rifles, and many others were employed as porters. Troops from South Africa and Rhodesia arrived, and together with the Nyasalanders, they formed the Nyasaland-Rhodesia Field Force. These forces, together with others from Kenya and Uganda, engaged in many conflicts against von Lettow-Vorbeck's soldiers throughout German East Africa and Portuguese East Africa. Throughout the war, troops, stores and vehicles passed through Nyasaland from the coast to the front line in German East Africa. An extension of the railway in 1915 to Chindio, where the Shire flows into the Zambezi, speeded the journey to some extent, but even so, the lines of communication were long and slow. In Nyasaland, vehicles were used as much as possible, and all the boats on the lake were commandeered for active service. However, the poor roads necessitated the extensive use of porters, who carried food and ammunition to soldiers on active service. The number of Nyasalanders who participated in the war effort is staggering. The *Handbook of Nyasaland* records that there were almost 19,000 men in the King's African Rifles, and around 200,000 men were recruited as porters and non-combatants.[55] The porters, or *tenga-tenga*, as they were called locally, numbered about 165,000, of whom about 2,600 were killed or died of disease.[56] Thus, the majority of able-bodied males in the country helped in one way or another. By the time the war ended, there was a considerable improvement in roads and communications; many of the motorcycles and lorries imported during the war remained in Nyasaland.

Wood's deployment to Cholo had a long-term effect on his career. Cholo is only about 25 miles north of Chiromo, but it is on the Shire Highlands above the eastern escarpment of the Rift Valley, and close to the huge massif of Mount Mulanje. It is an area of hills and valleys, with a cool, mild climate, and in Wood's day was covered in dense savanna forest, with evergreen montane forest in the valleys, along streams and on moist slopes. The rainfall in the Cholo area is one of the highest in Malaŵi. Planters had entered this area in the early days of settlement because of the rich soil and high rainfall, which made it particularly suitable for growing tobacco, tea and coffee. Wood evidently liked the Cholo district, as it provided relief from the very hot climate of the Lower Shire. His first references to Cholo were in September and October 1917.[57] The first of his birds from Cholo was collected on 14 September 1917, and he collected many others during the following weeks. And in October 1917, he wrote to Thomas: 'I am up at my place at Cholo for a week or two making bricks for my home, and during the last ten days over thirty small mammals have been brought in. I think I have at last found a man with some instinct for trapping things, fostered by a frequent disbursement of pence!'[58] One of the first small mammals, collected on 27 October in the hole of a mole-rat, was also the most special – a new species of Brush-furred Mouse, which was given the name *Uranomys woodi*.[59] By September 1919, Wood had left the cotton plantation at Chiromo and was living permanently at Cholo. 'My place' referred to a small estate which he called Magombwa (now incorporated within Kasembereke estate), a few miles south of the township of Cholo.[60]

Magombwa was set among steep-sided undulating hills, with bubbling streams flowing along the little valleys. In 1917, all the slopes were covered with savanna vegetation, and the valleys were thick with palms and montane forest. Damp leaf litter covered the soil in the valleys. It was cool and moist for much of the year, and mists hung over the forests in the mornings. Very few Africans lived here – it was too cold and moist for their liking. Wood's house was built on the top of one of the hills, with a brick and stone path leading up to it. The house itself was built of mud bricks with a corrugated iron roof.[61] Inside, it was relatively spacious, but small, with high ceilings and concrete floors, and with doors and windows made from local hardwoods. Along the southern side was a veranda, supported by brick columns and covered with a tin roof, and which looked out over the garden and the rolling hills towards Mount Mulanje. On one side, and adjoining the house, was a 'garden house', built in much the same style as the veranda, which in future years was to be incorporated into the house itself. There were no gutters (where temporary pools of water could breed mosquitoes), so rainwater flowed straight off the roof onto the ground. Outside was the kitchen house and a toilet house. It was not a large establishment, but quite adequate for Wood's simple needs and, importantly, it did not cost much to build. The steeply sloping garden was terraced, and in the valley below, Wood planted many trees and climbers that he had collected.[62]

Wood lived at Magombwa from 1919 to 1921, but nothing much is recorded about what he did there. The estate was planted with tobacco (as were most of the estates), and hence he must have been very busy supervising, growing and harvesting the tobacco – especially from about September (when the seedlings were nurtured in the nursery) to March (when the crop was harvested and dried). During these years, he collected hardly any specimens of birds and mammals, and his correspondence with Thomas came almost to a standstill, suggesting that the work of the estate kept him extremely busy all the time. However, his interests turned briefly to fish – a rather strange decision, since he was now living well away from the Shire and Ruo Rivers. In November 1919, he wrote to the Curator of Ichthyology at the British Museum of Natural History, Dr Tate Regan, requesting information about preservation of fish.[63] In the letter he mentions that he had 40% formalin and a small quantity of absolute alcohol, and says that he had experimented with various methods of preservation. He also wrote that he had been recently to Domira Bay on Lake Nyasa, where 'we had men fishing everyday with large seine nets', and told Regan that he wanted to make a collection of fish. As a result, Regan sent instructions for preserving fish, together with four cases of collecting equipment (tanks and jars), gallons of methylated spirit and the four volumes of Boulenger's *Catalogue of the Freshwater Fishes of Africa*.[64] In August 1920, a time when there was not much activity on the estate, Wood returned to Domira Bay and made a collection of 226 specimens that he sent back to Dr Regan.[65] This collection was, at the time, the most comprehensive collection from Lake Nyasa; it comprised 84 species, of which 46 were new to science (see Chapter 11). The other remarkable fact about the collection is that it showed that most of the fish in the lake were endemic (i.e. they occurred *only* in Lake Nyasa). Transporting preserved fish is difficult because of the weight of the preserving fluid. The British Museum of Natural History provided special boxes made of tin and wood; the size was determined so that when full of fish and fluid, each box would be a normal head load of 50 lbs in weight.[66] Hence, a collector could go to very remote lakes and rivers, from where the collection could be carried in such boxes to where it was shipped back to England. Regan acknowledged Wood's collection by naming two of the new species *Haplochromis woodi* and *Rhamphochromis woodi*. It is rather ironic that Wood's fish, which were amassed over a period of only a few days, proved (in subsequent years) to be one of his most significant collections (see Chapter 11).

The climate of the highlands must have been very beneficial to Wood's health, as by late 1920 he told Thomas that he was 'fit and flourishing'.[67] But the continual hard work necessary to run an estate, the many problems with finances and labour, and the uncertainty of crop prices were not really to Wood's liking. In December 1920, less than two years after he moved from Chiromo to Cholo, he decided to return to England. He wrote to Thomas:

> Am very busy packing up as I expect to leave here on a prolonged holiday in January. Shall arrive in England in May and will come and see you at once then. Will send home my mammal collection for Hinton [*another scientist at the British Museum*] to look through. . . . Now I've sold my estate here and as a result will have a very small income on which to live. I have certain obligations until end of 1922 and after that I shall be more or less free. So will discuss the possibilities of my doing some collecting work for you in N.E. of the Lake when I see you. I want to do the fish of the lake too if I can get funds.[68]

Wood's comments about selling his estate are strange because he resumed his life as a planter at Magombwa between 1925 and 1929. These remarks as a whole suggest a pattern that is evident throughout his life: he did not want to settle down in one place or be tied to permanent repetitive work. He was a nomad at heart and was unable to maintain an interest in something he did not like; he considered that jobs were only necessary to earn a modest income and to support his passion for collecting.

Wood left Nyasaland in early 1921. However, it seems that he did not go directly to England. Instead, he first went to the Seychelles islands, where he bought some land and a house (see Chapter 10). By August (or perhaps before), he had reached England, where, to begin with, he stayed with his father in Abercorn Place. But it was not long before he had embarked on a new career in North America, far away from Africa.

Chapter 5

SCOUTING AND PLANTING: 1922–1929

After Wood returned to London in early 1922, he visited the British Museum to see Dr Thomas, Dr Sclater and Dr Boulenger about his collections. He was still keen to continue his collecting (as he mentioned in his letters to Thomas), and was hoping that someone would finance the setting up of a natural history museum in central Africa and pay him to be the curator. However, this was not to be. Instead, Wood became involved with the Boy Scout Association, which had been founded a few years previously by Sir Robert Baden-Powell. Wood recorded that 'at Sir Robert's personal request, I left my estate and work in Nyasaland and spent three and a half years training men in woodcraft and campcraft'.[1] How Wood came to know Baden-Powell and secured this appointment is unrecorded, but undoubtedly Wood's outstanding physique and bearing and his experience in the African bush would have appealed to Baden-Powell. In late 1921 or early 1922, Wood was appointed as Assistant Camp Chief for Overseas, and, in the first months of his appointment, he held camps in Rhodesia and Cape Province.[2]

Much of Wood's time in the Scouts was spent in Canada where, in 1922 and 1923, he ran training camps throughout the whole of the Dominion. In early July 1922, he was the 'Assistant Camp Chief' for the first Canadian Gilwell Camp in Ottawa, where he trained Canadian Scouters. Scouting was becoming popular in Canada, and many boys were becoming Wolf Cubs or Boy Scouts. Baden-Powell's view was that training boys to become self-reliant, proficient in survival skills, and helpful and considerate to others were essential prerequisites for the making of healthy, happy and useful citizens. Wood's role was to train scouters who would then encourage more boys to become cubs or scouts, and hence to fulfil the aims of the scout movement. In 1923, when Sir Robert visited Canada, he said that he was delighted to be able to send Rodney Wood to help them train their scouters. Canada was evidently very pleased with Wood and his work, and in late 1922 he was given the title of Dominion Camp Chief.[3]

The year 1923 was a very busy one for Wood. After spending the first part of it in Canada, he was asked to return to England for the summer to become Camp Chief at Gilwell Park near Chingford on the edge of Epping Forest, near London. Here he ran courses for the Wood Badge. The term 'Wood' had nothing to do with Rodney Wood, but was associated with woodcraft – the ability to live and survive in woodlands or 'bush'. A recipient of the Wood Badge was entitled to wear a special bead hung on a leather thong around the neck. The symbolism of the badge is related to Baden-Powell's experiences in the Boer War, when he had to rely on his bushcraft skills. The beads were originally from a necklace worn by a Zulu chief whom Baden-Powell captured. The leather thong is a reminder of the time when an old African man saw that Baden-Powell was downcast, and to cheer him up, the old man gave him the leather thong from around his neck and told him: 'Wear this; my mother put it on me for luck; now it will bring you luck.'[4]

The first Wood Badge course was held in 1919.[5] The syllabus covered many aspects of woodcraft, all of which were very familiar to Rodney Wood.[6] The syllabus included:

1. Troop organisation: drills, formation, scout's pace, ceremonies, flag raising, physical exercise.
2. Campcraft: camp selection, tents, camp fires, cooking, first aid, sanitation.
3. Fieldwork: distance measurement, mapping, drawing, use of compasses, stars, night marching, report writing.
4. Pioneering: axemanship, construction of bridges, huts and shelters, use of tackle.
5. Woodcraft: observation and deduction, nature notes; identification of birds, animals, and trees.
6. Signcraft: signalling (hand, whistle, smoke, semaphore and morse), nature trails, sand-tracking, weather lore (clouds, etc.).

7. Pathfinding: patrols (with sealed order for 10-hour journey in Epping Forest), leaf-collecting, report (including sketch map of trek), panoramic drawings from given points, etc.
8. Camp games.

One of the scouters who assisted Wood at Gilwell Park was Don Potter, who worked at Gilwell Park for ten years as a troop leader in the 1920s. Subsequently he became a professional sculptor and taught at Bryanston School in Dorset. Seventy years later (in 1995), Potter was able to recall Wood and those far-away days at Gilwell:

Of course I well remember Rodney Wood who made his appearance at Gilwell Park in the early 1920s. I chiefly remember his imposing personality, tall and handsome, a very good speaker. He had been living in the Seychelle Islands, was educated at Harrow, and his home was somewhere near there. He was a great naturalist. I remember one occasion when he took us to his home to see his collection of trophies and stuffed animals which he had hunted.

He got on well with everybody with his easy friendly manner. I was young at the time and very much admired him, in fact he was my hero! as I'm sure he was to many of us there.

He was about 6–10 years older than me as far as I remember, and like so many of us, going through the stage of searching for our souls, and future occupations in life.

In our spare time we practised archery, lassooing, rope spinning and throwing the boomerang. Rodney joined us in all these activities.[7]

One of the participants on the Wood Badge course in May 1923 was Arthur Westrop. Although neither he nor Wood could know it then, their meeting at that time was to have an immense effect on both their lives in future years. Arthur Westrop was born on 30 August 1893, four years after Wood was born. He attended Harper Adams Agricultural College and later won a scholarship to the University of London, where he studied entomology. At the outbreak of war, he joined the Horse Artillery and was involved in the offensive at Mons in 1917, where he was wounded and awarded the Military Cross and the Croix de Guerre.[8] After the war, he turned his attention to agricultural chemistry, and in

1920 he joined a company which had extensive agricultural interests in tea and rubber in Ceylon. Westrop was a scoutmaster in Ceylon, and on his first leave in 1923 he decided to enrol for the Wood Badge Course at Gilwell Park. He was an outdoor man and a keen naturalist, and so had many interests in common with Wood. Many years later, Westrop recalled how his friendship with Wood began:

Captain Gidney, Gilwell's first Camp Chief, was away . . . and his place had been filled by a volunteer, Rodney Wood. The Acting Camp Chief, a commanding figure over six feet in height with piercing blue eyes, was said to be a Big-game hunter from Africa, who had hunted lions with a bow and arrow. He was, at that moment giving a talk on birds, a subject in which I had been greatly interested since boyhood. . . . This was a talk by a man who was an expert on the subject, knew the songs, food, habits, nests and eggs of birds, described methods of identification which did not involve killing them, obviously a man who loved them. Wood related a story of Wagtails on his tobacco estate in Nyasaland and, in response to a question by a member of the course, gave a brief description of the whereabouts of the country.

The letter box at Gilwell was then a hollow log, but a bird had built a nest in it and was feeding its young. A notice requested that, for the time being, letters should be posted in a soap box nearby. Before closing his talk, Wood had drawn attention to this notice saying that a Great Tit was occupying the log.

Now I had already watched the parent birds popping in and out of their nest and was sure that they were Blue, not Great, Tits. Not for nothing had my boyhood been spent in the country, where a letter-box, placed at the entrance to the drive leading to my home, had annually been the nesting site of one or other of these two species. . . . I knew them both well. Such a mistake in identification could not be allowed to pass. Greatly daring, I stood up. 'Please Sir, it is not a Great Tit, it's a Blue Tit', I said.

The Camp Chief paused. It was unusual for a member of the course to bluntly contradict its Leader. . . . But his own experience of birds had been gained largely in Africa . . . Perhaps this interrupter . . . might be right. 'Those of you who have fieldglasses, go and get them, the class will adjourn to the letter-box to identify the birds', the Camp Chief ordered. Everyone adjourned to the letter-box where the bird was correctly identified as a Blue Tit – 'a verdict accepted without comment by the Camp Chief'.[9]

After the course, Wood and Westrop went their separate ways. Westrop returned to Ceylon, where he continued his work on tea estates and became the first Deputy Camp Chief of Scouts in Ceylon, and Wood organised further Wood Badge Courses at Gilwell Park before returning to Canada in the autumn of 1924. During the winter months he made frequent trips from Ottawa to Montreal and Toronto to run Gilwell Winter Training Courses, and during the summer he was Camp Chief at courses in Ottawa, Alberta and British Columbia. At the conclusion of the camp in British Columbia, Wood travelled (as the Canadian representative) to a conference in Colorado organised by the Boy Scouts of America.[10] Although details are sketchy, Wood seems to have led a very busy life in Canada, but he also had time to learn about game conservation with the Dominion Parks Board, and to write a booklet entitled *Animal Tracking for Boy Scouts*.[11] He also became involved with Red Indians in Canada, and it was said that he was inducted into one of their tribes; at times he wore traditional Indian dress. He resigned at the end of 1924 – maybe he had had enough of cold winter weather, or his longing for the Seychelles and Nyasaland finally got the upper hand. In the Annual Report for 1924, the Boy Scouts Association recorded their appreciation of his services:

> At the end of November, Mr Wood resigned as Dominion Camp Chief and left for the United States where he intended to remain for a short time, to act as Camp Chief at a number of training camps, thereafter returning to his home in the Seychelles Islands. At its meeting in November, the Executive Committee placed on record a resolution of appreciation of the fine Scout spirit and eminent qualifications which were characteristic of Mr Wood during his work in Canada in his capacity of Dominion Camp Chief.[12]

Realising that he would have to change boats in Ceylon on his way from the United States to the Seychelles, Wood wrote to Arthur Westrop (whose address he must have kept since the Gilwell days) to ask if he could visit him for a few days. The relationship blossomed during this visit, as they were no longer teacher and pupil, but fellow agriculturists, scouters and naturalists. As Westrop recorded: 'Rodney, with whom we were soon on terms which permitted the use of Christian names, a privilege less common than it is today, proved a most interesting guest. He soon revealed a knowledge of natural history, botany and kindred sciences possibly unrivalled in a single person.'[13] Westrop went on to extol Wood's love and immense knowledge of birds, butterflies, fish, shells, trees and orchids. The two men got on famously. Wood also told Westrop about the problems of farming in Nyasaland, fluctuating prices, the fickle nature of the rains and the difficulties of employing local labour on the estate. Wood looked into the possibility of planting tea in addition to tobacco on Magombwa, and into the ways that Ceylon could help with the problems of Nyasaland. The days passed far too rapidly, but Wood had instilled in Westrop a fascination for Nyasaland; the outcome was that, many years later, Westrop became the owner of Magombwa and spent the last twenty years of his life in Nyasaland.

Wood arrived in the Seychelles in mid-1925 and stayed until November of that year.[14] His life in the Seychelles – the subject of another story (see Chapter 10) – was totally different to that in Nyasaland or Canada. In the 1920s the Seychelles were very remote from the rest of the world, and in the soft, warm tropical climate he led a life of tranquillity and peacefulness. During this period of five to six months, while building a little house for himself, Wood was offered the post of Camp Chief and Secretary of Scouts in the Gold Coast, and he also applied for a post in the Colonial Service. However, he did not take up the post in the Gold Coast, and nothing became of his application to the Colonial Service, for he left the Seychelles on 18 November for Mombasa, Zanzibar and Beira, and eventually arrived at Luchenza (the nearest railway station to Magombwa) on 1 December. It was a rotten journey because, for the whole time, his body ached with pain and he had a violent headache, both of which he assumed were symptoms of Dengue Fever.[15]

During Wood's absence, Magombwa had been looked after by a neighbour, Mr McLeod. On his return, Wood found the estate in moderately good condition, and the three 'capitaos' (senior African staff who looked after everyday running of various parts of the estate) were still there. He commented in his diary that everything had grown beyond belief and that the estate was looking very beautiful. Changes had been made to the house, making it more spacious and pleasant to live in. It was December, and the wet season was about to begin, so the young tobacco seedlings had to be planted in the fields. Wood thus quickly resumed his life as a planter at Magombwa.

The Cholo area was changing rapidly as new estates were established, more European planters were settling, and more Africans were being hired as labourers by the estates. Slowly, the face of the countryside was changing from the natural *Brachystegia* woodland to large fields with remnant patches of forest only in the gullies and on steep slopes. The roads were improving (although they were still poor in many places), and most planters had some form of motorcycle, car or lorry so that travel was much easier and quicker than it had been prior to the war. Although Wood lived on Magombwa, rather isolated from other planters, he had many friends and acquaintances with whom he met frequently. Wood's diary gives some idea of the day-to-day life at Magombwa; for example, in early February 1929:

31.i.29 Holing trenches for plants on D3 garden which is now bunded and ridged . . . Took male and female P. bromias on bougainvillea, and saw two more males on zinnias.

2.ii.29. Work on extension of seedbearers for plot D4 all morning. Office after lunch, over to Mwalampanda for night after tea.

3.ii.29. (Sunday): Back after breakfast. Mr and Mrs Lee and Mrs Matthews came in for half an hour about 11 a.m.; gave them mangoes. Robert and Margaret [Harper] over to lunch and tea; went to inspect new tea garden, D3, with R.

4.ii. 29. Work on D3 and started planting out seed at stake there. Bookwork in afternoon.

5.ii. 29. Terracing seed-bearers field D4. Personal work needed all the time. Planting out seed. A tremendous swarm of 'lintumbu' warrior ants appeared all round front of house at tea time. The ground swarming with them everywhere. They killed a centipede by flowerbed and insects of all kinds. Shot a Buzzard Eagle (Kampifalco monogrammica), stomach containing remains of a young chicken and one lizard. Was sorry to kill it but chickens have suffered a bit of late.

6.ii.29. In to Blantyre for provisions and to bring bring out Jay Williams who is staying with me for a bit. Spent night with Wrattens.

7.ii 29 Back to Magombwa.

8/9.ii.29. Work on tea gardens all day.

10.ii.29. (Sunday) To Zomba to stay with two days with Hawkins and give his KAR officers a bit of bushcraft training.

11.ii.29. Zomba. Lecturing and demonstrating on animal spooring, human spooring, knotting, odd bushcraft all day. H.E. The Governor attended the human spooring session, and we dined with him that evening. Songs and mirth at GH and a most pleasant evening.

12.ii.29. A short lecture on animal habits first thing and then back to Blantyre where sundry business, and stayed night with Wrattens again.

13.ii.29. Back to Magombwa in morning, taking Archie out with me again as he was unable to get back to school owing to washouts on line.

14.ii.29. Magombwa. At work all morning on terracing seed bearer gardens, and planting seed at stake in D3. Office and posho [*food*] in afternoon.[16]

These brief entries give some idea of the chores involved in running an estate. Constant supervision was essential to ensure that planting, hoeing, picking and all the other procedures of tobacco and tea production were done correctly. There was a continual stream of work in the office keeping track of accounts, labour records, orders and sales. Pests of various sorts – warrior ants, predatory birds and baboons – had to be controlled. The weather and the state of the roads determined what could be done each day, and they were a constant source of comment by Wood – for example, 'fine all day', 'rain continues heavily daily', 'weather continues very wet', 'road awful', and so on. Some of this tedium was relieved when visitors came to Magombwa, or when Wood visited his friends and neighbours, often staying for meals and overnight. There were also lighter and more enjoyable times, such as visits to Zomba, when Wood was able to indulge in his love of partaking in interesting conversation and listening to songs. Wood regularly visited Blantyre, some 25 miles away, to collect provisions. In this sort of life, flexibility was essential because each day brought its own problems – one never knew when rain, washouts, impassable roads, pests, labour problems, lack of supplies and myriad other eventualities would overturn the best-laid plans.

On his return from Ceylon, Wood started planting tea on Magombwa.

In Ceylon, he had been greatly impressed by the way in which fields on slopes were contoured in order to avoid soil erosion. This method is called the 'reversed slope' method, or 'reverse bunding', and it was developed by a tea planter called Denham Till.[17] It involves pegging the planting holes in the original soil exactly on the contour, and then excavating the soil from the hillside above the plant to form a terrace so that the rear is slightly lower, and the front is slightly higher, than the planting point.[18] The excavation is continued along the planting line so that a continuous terrace is formed. The advantage of reverse bunding is that the terrace is able to hold the water after a heavy rainfall without spilling over the front edge; as a result, the water soaks into the terrace, soil erosion is minimised or completely stopped, and the new plant takes root in the original soil profile. Wood was responsible for introducing reverse bunding into Nyasaland, and now it is used extensively all over the country for virtually all crops. However, it took many years to convince people, especially Africans, of the value of reverse bunding. The traditional method of planting crops was up and down the slope, a method which allows water to flow down the slope, taking all the topsoil with it; as a consequence, there are many places where the depth of the soil is much reduced from its original level or, at worst, where the topsoil has been totally removed. The beneficial value of reverse bunding, in terms of increased crop yield and conservation of the soil, is incalculable.

Wood was not the first to plant tea in Nyasaland. The first tea plants were imported into the Protectorate in the 1880s and were planted in the grounds of the Blantyre Mission. These plantings were experimental in nature, but they did prove that tea could survive in Nyasaland. However, commercial tea growing required a damper climate than existed at Blantyre, so it was in the south and southeast of Mount Mlanje that the first tea plantations were established. The credit for establishing the first commercial tea estate is usually given to Mr and Mrs Henry Brown who, before coming to Nyasaland in 1891, were coffee planters in Ceylon.[19] In the early years, coffee was the most common and widespread crop in Nyasaland (besides tobacco), but during and after the First World War, the value of coffee declined and production decreased. Lauderdale and Thornwood Estates, where the first tea had been planted, gradually replaced their coffee bushes with tea bushes, and over the course of the following decades, all the estates in the Mulanje and Cholo areas replaced both their coffee and their tobacco with tea. As a result, tea production increased over the years, as did the amount of tea exported annually: 43,000 lbs in 1912, 155,000 lbs in 1918, 801,000 lbs in 1920, 1,426,000 lbs in 1928 and 1,939,000 lbs in 1930.[20] It is hardly surprising that Arthur Westrop used the term 'Green Gold' to describe the crop! This growth in production is all the more surprising considering the very small area in Nyasaland that is climatically suitable for tea – only around Mulanje and Cholo in the southeast of the country, and near Nkata Bay on the northwestern shore of Lake Nyasa. The increase was due primarily to new strains of tea being imported from India and Ceylon, and to an increase in the amount of land devoted to tea – rising from about 2,600 acres in 1912 to nearly 10,000 acres in 1930.[21]

Perhaps one of the most delightful times for Wood was when Arthur Westrop visited Magombwa in 1927. Westrop had been on leave in England and was returning to Ceylon. Sometime between 1925 and 1927, he had invested some of his savings in Magombwa as a way of helping Wood to establish tea at Magombwa and to increase the acreage under production. Wood must have painted a wonderful picture of Nyasaland to Westrop, extolling all the virtues of the country and his estate – so much so that Westrop came to think it might be a suitable country for him and his family to move to in the future. Wood had enumerated all the types of fruit that he grew on his estate, including banana, papaya, avocado, mango, pineapple, passion fruit, orange, lime, lemon, grapefruit, pomelo, peach, loquat, granadilla, rose apple and tree tomato, as well as two fruits – mulberry and strawberry – not normally associated with the tropics.[22] To get to Nyasaland, Westrop travelled on the Union Castle Line boat *Grantully Castle* to Cape Town, and then caught the train from Cape Town to Salisbury, which took four days, and then on to Beira. Since the Salisbury–Beira train and the upcountry train to Nyasaland were not synchronised, he had to wait for three days in Beira, which he considered was a rather 'primitive' place in those days. In 1927, the railway line in Portuguese East Africa ran from Beira to Murraco on the southern bank of the Zambesi. Here, Westrop and the other passengers transferred to a ferry boat which spent the night zigzagging across the river trying to avoid the sandbanks. At Chindio, on the north bank of the Zambesi, he boarded the Shire Highlands

Railway; the first stop was at Port Herald (for customs and passport control), then on to Chiromo, and finally Luchenza. Wood met Westrop at the station in a Ford box-body car (a very popular model in Nyasaland) and they drove to Magombwa. This was Westrop's first visit to Africa, and so everything was fascinating to him. Together, they walked all over the estate, looked at the new tea that had been planted, investigated possible sites for a small hydro-electrical power station on the Nswadzi River and for a tea factory on one of the hills, and talked to Wood's neighbours about the future of tea on Nyasaland. The visit was far too short – only a few days – but was sufficient to show Westrop that he was happy with his investment in Magombwa, although he considered the estate was not large enough to be a viable proposition if prices slumped. Before leaving Nyasaland, Westrop drove to Blantyre and was impressed by the almost unbroken canopy of trees that hid the distant views of the hills.

The finances of an estate were (and still are) greatly affected by fluctuations in the price of crops. A planter's worst nightmare is the failure of his crop because of inclement weather, and a dramatic fall in the price of the crop because of forces beyond his control. Magombwa was only just viable as an economic plantation. In the late 1920s, most of the revenue came from tobacco. The year 1927 was a particularly good year for tobacco, and planters received an excellent price of about 2/- per pound. Wood and many other planters increased their acreage for tobacco in 1928 in order to capitalise on this good price. However, in 1928, over-production and difficulties of blending Nyasaland tobacco with other tobaccos caused the price to fall to just 6d per pound; much of the tobacco was not sold until years later.[23] At the same time, the preparation and planting of the tea was expensive, and it took several years before a crop would be able to produce a profit for an estate. The fall in tobacco prices meant that there were inadequate profits to pay Wood's salary, and it also caused a drop in the value of land at Cholo. When Westrop visited Magombwa in 1927, he realised that the area of the estate should be increased and he attempted to buy some adjoining land. When this proved to be impossible, he decided to invest more of his own funds in Magombwa, and by 1929 he had become the sole owner of the estate. By this time, Wood had accepted the post of Game Warden (see chapter 6) and so Westrop appointed one of Wood's close friends at Cholo, Robert Harper, as the manager. Westrop decided never to plant tobacco again because of the instability of the market; instead, he concentrated on tea, which he knew and understood well and which was well suited to the climatic conditions of Cholo.

When Wood returned to Nyasaland in 1925, he started to collect again even though running Magombwa occupied much of his time. His field notebook for his bird collections shows that he collected only six specimens in 1926, all of them at Cholo, or, more likely, at Magombwa.[24] But in 1928 he managed to get away from Magombwa for several collecting trips during the dry season, after the tobacco had been harvested in March–April and before the planting of the new tobacco in November–December. In May and June of that year, he visited the Bua and Ludzi rivers (between Lilongwe and Kasungu); in September, October and November he collected at several localities on the lower Shire; and in early 1929 he collected a few specimens at Cholo.[25] During this entire period (1926–1929), he collected about 100 bird specimens but hardly any mammals – just a few mongooses and palm civets from the estate.[26] In contrast, Wood collected butterflies at every opportunity. Butterflies are much easier to collect than vertebrates: in the field, the collector needs just a net and a killing bottle (as well as a lot of energy and determination to catch the fast-flying species), and at home all that is needed are special entomological pins, setting boards and good pest-proof storage cabinets. During his years at Magombwa, Wood collected around the estate (although all labels state 'Cholo' as the locality) and at Cholo Mountain, Blantyre, Zomba Mountain, Chiromo, Port Herald, Fort Manning, and on the Bua River. These localities (and the dates of collection) are mostly the same as where he collected birds. As well as collecting, Wood made careful observations about the butterflies in his garden. He published some of these observations in two letters to the Entomological Society of London. Both letters illustrate his powers of observation and his attention to detail, and also how the quest for information about butterflies was always uppermost in his mind. His first letter, published in 1927, described the crepuscular dances of *Libythea* butterflies. After writing about dances of moths which he had observed in Canada, he continued:

On January 1st 1927, I was sitting on my verandah in the Cholo district of Nyasaland when I noticed a specimen of *Libythea laius* pass over the house and fly down the open park-like garden in front, checking at and flying round several low trees or bushes on its way. Grabbing a net, I followed, as I was anxious to take a specimen, many having passed over that afternoon in rapid sustained flight towards the east, usually rather high in the air. . . . . I lost sight of the butterfly by a *Juniperus procera* tree about twelve feet in height touching a somewhat larger wide-spreading native tree, so stood still looking around to see if it would show again. I then became aware of a dark insect which at first I took to be a moth, flying in wavy undulations round about these two trees. Noticing it continue, backwards and forwards round and round in a seeming kind of gambolling flight, I made a sweep at it with the net, but it dodged and at once settled motionless in a sprig of the juniper. This instant dash to cover and subsequent 'freezing' is a marked habit of many of our neutral-coloured butterflies in this country when threatened with any danger. Going up quietly I found it just level with my face and was astonished to see it was the *Libythea* I had been following. . . . For fully another ten minutes it kept up its rapid gambols or gyrations round the tree. I had shot after shot at it in the failing light, as it came regularly on the same line of flight, but its speed and dodging powers were astonishing, until at last I lost it completely in the gathering darkness.

Here I may remark on a very curious phenomenon of *Libythea* in this country. Each year about the first or second week in January enormous numbers of the butterflies appear for a day or two, invariably flying in one direction, though as far as I can ascertain at present this direction is not constant. In January 1926, Mr C. F. Belcher told me that while motoring from Zomba to Blantyre he had met swarms of thousands of butterflies for the whole distance of some forty miles. . . . They were then flying in a northerly direction. A day or so later not a single one can be found. They will not appear again till the following year.[27]

The second letter concerns his fascination with injuries inflicted on the wings of butterflies by vertebrates:

I planted a large bed of Zinnias which prove attractive to certain species. . . . Such masses of butterflies congregate here that they are always in a continual state of disturbing each other, and fly to such trees as mangos, local large Brachestegias, Bougainvillea-covered trees, etc. which surround the flowerpatch. On these trees are great numbers of small lizards, which I fancy account for most of the damage at the moment when the butterflies settle on these trees. Once settled unobserved, I have not seen one attacked yet. Many species of insect-eating birds are present such as *Batis molitor*, other flycatchers, many species of shrikes . . . For the last three weeks, I have been watching one of the really big chameleons literally lapping up butterflies from the Zinnias. It sits on a small branch of Bougainvillea just over some of the Zinnia flowers and takes each butterfly that comes along with unerring marksmanship. I am trying to find out how often insects escape from it, but so far without success. It takes *Papilio lycaeus* which swarms here daily, with great ease. So I am forced to the conclusion that it is the smaller lizards which are responsible for most of the wing injuries. All our geckoes seem only to take anything while it is actually moving, no matter how slightly . . . whereas if completely still they appear to lose all interest in it, even after they have stalked it while it did move. . . . Nearly every species of *Charaxes* I take has suffered to its tails, as *Charaxes* nearly always 'oscilate' their hindwings if any danger approaches them. Papilios, of course, vibrate their wings all the time they are feeding on the flowers, and the smaller lizards are able to dart at them from branches in contact [with the flowers] or even from the stalks of the zinnias themselves. . . . I am getting to associate certain irregularities of the wing outline with the possibility of either lizard or bird damage . . . What chiefly amazes me is the ability of butterflies to carry on a very satisfactory flight after literally appalling damage to all wings, provided the costal area of the fore-wings is moderately untouched.[28]

With this letter, Wood sent many specimens to illustrate the damage caused by the lizards and birds.

After his return to Nyasaland in 1925, Wood took an increasing interest in the affairs of Nyasaland, especially in connection with land use and conservation. Although he had no formal training in ecology (a term that was unknown to most people in those days), his vast knowledge of natural history and his keen powers of observation enabled him to interpret changes in the environment that went mostly unnoticed by other people. Like most Europeans in Nyasaland, Wood read the *Nyasaland Times* which, as well as providing news, engaged in lively debates on many issues pertinent to the life and functioning of the colony. When it came to issues of natural history, Wood could not resist adding his contribution to a debate. On 16 March 1926, Wood noticed a short paragraph headed 'The Exception', which included the following lines:

A discovery that guinea fowl and other game birds are valuable allies of the cotton grower has been made as a result of experience gained by a South African farmer last season. Apparently, on account of heavy rains, the cotton crops . . . were generally poor. The growers, with one exception, barely cleared expenses. One planter, however, who farmed under exactly the same conditions as the others had a bumper crop. Two hundred wild guinea fowl flocked following his plough and cultivators and cleaned away insect pests. It is believed by growers in the district that the presence of those birds accounted for the success of this land. . . . If complete flocks of wild and tame birds can be ushered to cotton lands, immediately improved crops should be the result. Farmers in general instead of shooting off game birds should make sanctuaries for them, and great benefits (this grower believes) would be reaped thereby.

This article prompted a reply by Wood which shows that, if necessary, he had a subtle sense of humour and could be quite sarcastic, but also that he understood the role and importance of animals in nature. Under the heading '"The Exception" – and a pair of Wagtails', he wrote:

Sir, In your issue of 16 March, I notice a short paragraph entitled 'The Exception'. It seemed to me that the article that followed was being quoted by you and was not in your own words, but that is beside the point. I was amazed beyond conception by the fact that guinea-fowl and other game birds were beneficial to crops apparently had only just been found out in South Africa. . . . If that is the true state of nature knowledge in the Union, then all I can say is, may the Almighty God help them, for they will sure need that help! 'The Exception' – exception to what? To the idea that all birds are man's mortal foes? I could not fathom it.[29]

After pointing out that Canada and the USA have signed an act to protect migratory birds, and that there is a monument in Los Angeles to the honour of the 'birds that helped to win the war', Wood continued:

It seems beyond belief that such ignorance on the economic position of birds can exist in the twentieth century, and among a so-called 'highly civilised' people. Of course, it is the whole effete system of education that is to blame. Nature study in its widest sense, its economic relationship to man's life and needs, is not taught . . . intelligently or practically. When it has dawned on those responsible how great the need is, let us hope it will not be too late.

In the face of climatic vagaries, insect pests, fungoid diseases, and high taxation, I try to grow tobacco. On one of the few occasions when it actually gave me profit, I owed every leaf to a pair of wagtails. In the unlikely event of my ever becoming rich, I shall erect a statue to them. This season, I hoed up a five-acre field and shuddered at the magnificent fertility of the cutworm as evidence by its numbers. I did not see young tobacco plants having much chance there. But a flock of our common crows, up to sixty strong, took the matter in hand. They worked that field systematically just before I planted it out, and I gave them every facility to carry on. There is tobacco in that field now *and no cutworm*. It should be well-known in these days that only a very few birds are harmful to man's interests. In each country they differ slightly owing to local conditions. A crow may be mightily hard on groundnuts out here, but most of us grow tobacco, cotton or maize. A few groundnuts are a very cheap *insurance premium* for cutworm destruction. *Birds are the cheapest and most efficient 'insecticide' known*. An odd bit of fruit or a few maize seeds are a cheap premium, as also is the odd chicken, for vermin destruction en masse. If bird sanctuaries and intelligent bird encouragement on estates were more common, so also would be the returns from our crops I am, etc.,

RODNEY C. WOOD [Wood's italics]

Sadly, Wood did not say exactly how the wagtails protected the tobacco leaves! His vigorous fight for the protection of birds was evident in his next contribution to the *Nyasaland Times* on 28 May 1926, on 'Gun-Boys'. Gun-boys were Africans who were stationed in the fields to harass and shoot birds that were regarded as pests. This was normal practice for most estate owners, although some (including Wood) wanted to legislate against their use. He wrote:

Sir, At two meetings at which I was present recently, opinion was expressed strongly in favour of the Government enacting legislation to abolish the 'gun-boys'. It would to be very difficult to make out a valid case for permitting their retention in face of the known damage and destruction they cause. Any agriculturalist who has studied the question knows the inestimable value of game birds on his land. Our economic crops are a prey to cutworm, plant-bugs, grasshoppers, etc., and great losses are occasioned thereby.

Before I had realised the enormous amount of good they do, I myself used to shoot francolin and guineafowl and quail on my former maize and cotton plantations. I then made a systematic examination of the stomach contents of birds shot. There is no need for me to go into details of results, but the statement that often well over two hundred injurious plant-bugs would be found at any one time in one stomach may give some idea of the benefit conferred by these birds. Quail are one of the great destroyers of weed seeds known. But they do not enter into this gun-boy question; they are happily too small a mark. These gunboys trespass over private estates and shoot anything they can find as often as they think they can do it with impunity. Land owners who are desirous of protecting their game-birds are caused much justifiable annoyance . . .[30]

There was considerable debate at this time about the value of gun-boys; the opposite view was that there should not be a ban because game birds provided food for both Africans and Europeans. However, the government did legislate against gun-boys in late 1926, and the Cholo Planters Association felt that it was able to take some credit for the new Act: 'Planters generally have viewed with alarm the rapidly disappearing bird life in the Protectorate. Largely owing to the efforts of the Association, backed up by the N.P.A., and ably supported by the logical and forcible propaganda of Mr Rodney C. Wood, the government have been induced to announce that "no further licences to bird boys will be issued and that at the expiration of existing licences, such licences will not be renewed"'.[31]

Although Wood must have enjoyed his life at Magombwa, it seems that he did not want to spend his whole life as a planter. Magombwa was certainly a lovely place to live, he had good friends, he had the chance to travel around southern Nyasaland, and he was able to indulge in his love of collecting. However, the financial uncertainty of running an estate, and his lack of private capital, must have weighed heavily on his mind. In addition, his love of the bush and wildlife encouraged him to look for employment where he could utilise his wide knowledge and experience in wildlife conservation. So, on 2 February 1928, when the tobacco crop was reaching a bumper harvest, he applied for an appointment in the Colonial Service as 'Game Warden, Nyasaland or, failing that, Assistant Game Warden in any other tropical African country'.[32] Wood suggested that his salary should start at £360 per annum for the first two years, rising to £400, and later to £600. By this time, he knew many officials in the government, and was well known for his strong commitment to conservation. However, there was as yet no such post of 'Game Warden' in Nyasaland, and it must have been Wood's hope that such a post would be created just for him. It took about one year for all the formalities to be completed, partly because all correspondence had to be passed from the colonial government to the Colonial Office in Downing Street. In May 1929, however, Wood was finally offered the post of 'Game, Cultivation and Tsetse Fly Control Officer' (soon abbreviated to 'Game Warden') at a salary of £480 per annum, rising in £20–£30 increments to £720 – and he accepted immediately. So he left Magombwa in the care of Robert Harper, and started his new life as the first Game Warden of Nyasaland.

Chapter 6

HUNTING AND GAME WARDEN

At the time when European explorers first visited the country that was to become Nyasaland, game was extremely abundant. One of the first accounts was given by Henry Faulkner who, in 1867, travelled up the Shire Valley with Lieutenant Young to look for David Livingstone who, it was feared, had been killed by Ngoni warriors. At this time, Europeans had not settled in this part of Africa, the slave trade was still flourishing, and the establishment of the Protectorate of British Central Africa was 25 years into the future. Faulkner was one of the old-style hunters who shot indiscriminately at anything large (and hence much of his book makes rather unsavoury reading). However, he also included comments about the great herds of large mammals that lived on the plains of the Upper Shire Valley and close to the lake. For example: 'I saw a large herd of zebra . . . The beautiful creatures were grazing in an open space in the sun.'; 'I found myself traversing an extensive plain and the waterbuck, pallahs [= impalas], reedbock and gazelles might be seen wherever I looked, as far as the eye could reach. I never saw such quantities of game anywhere.' A little later, when he came to the river's edge and climbed to the top of a large termite mound, 'everywhere, as I looked over the plain, herds of antelope were to be seen. The glasses now showed me that there were many hartebeest among their number.' On another occasion, he wrote 'Koodoo, pallahs, hartebeest, gemsbok and reedbock: wherever the eye turned antelopes of some kind were to be seen.' In addition, there were lots of buffalo and elephants all along the banks of the Shire River. Faulkner's conclusion was: 'I do not believe there is another such field for sportsmen . . . as the banks of the upper Shire.'[1] A few years later, Drummond[2] was equally impressed by the animals that he saw: 'The waters of (Lake) Shirwa are brackish to the taste and undrinkable; but the saltiness must have a peculiar charm for game, for nowhere else in Africa did I see such splendid herds of large animals as here. The zebra was particularly abundant . . .'

By the last decade of the nineteenth century, many Europeans had settled in the Protectorate, mostly in the south. By this time, the numbers of some species were already in decline as a result of land clearance for agriculture and hunting. Elephants in particular had been over-exploited for their ivory. In 1891 and 1893, for example, the official figures for the export of ivory show that it was by far the most impor-tant export from British Central Africa,[3] and by 1897, elephants were becoming rare in much of the Protectorate.[4] Hunting was a popular pastime amongst colonial administrators, army officers and planters. Maugham, for example, went on weekend hunting trips in 1894 near Matati on the Phalombe Plain, a few miles from Zomba. In this region, he and his friends were able to shoot impala, eland, bushbuck, duiker, sable, Lichtenstein's hartebeest and warthogs, and they observed foot-prints of elephant, rhinoceros and lion.[5] Another colonial officer, H. L. Duff, also provided contemporary accounts of wild animals while he lived at Zomba from 1898 to 1902.[6] He, like Johnston, was an obser-vant naturalist and recorded natural history notes about insects, reptiles, birds, and mammals. Among his interesting observations are: 'The rhinoc-eros, though scarcely common, is to be met occasionally'; 'Hippos swarm in all the considerable rivers and lakes where they are subjected to continual persecution . . .'; 'The buffalo may be found without much difficulty . . . great herds frequent the Elephant Marsh . . .'; 'Lions and leopards are plentiful all over British Central Africa'; and 'Herds of zebra are everywhere common . . .' Duff also comments on some of the antelopes: 'Sable are still abundant in Central Africa – the Shire Highlands especially are full of them'; 'The eland is fairly common on the Shire Highlands, in the valley of the Shire river, in the Henga valley and in the neighbourhood of Deep Bay, and more or less over the Protectorate'; 'The hartebeest is by far the commonest antelope to be found in British Central Africa . . . and is killed in considerable numbers by both mankind and by beasts of prey.' Duff recorded that many of the other antelopes – waterbuck, impala, reedbuck, bushbuck, impala, oribi, duiker – are common in most parts of the Protectorate.

In spite of the intense hunting of the nineteenth century, game was

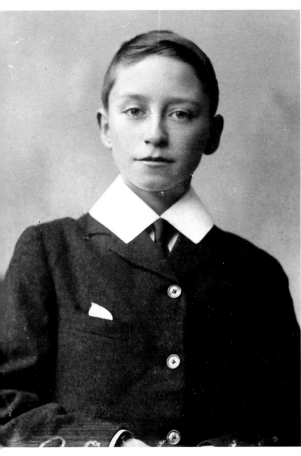

Left:
Rodney Wood in 1903, aged 13,
during his first year at Harrow
(Harrow School Archives).

Left:
Rodney Wood with his
father, Alexander John
Wood, about 1906.

Four early naturalists in Nyasaland.

Clockwise from top left:
Sir Harry Johnston; Alexander Whyte; Sir Alfred Sharpe; Sir John Kirk.

above left: Rodney Wood at either Chikonje or Magombwa, circa 1914-1920 (Historical Society of Malawi Archives).

above right: Monkey Bay, 1945-1947 (Rosemary Lowe-McConnell).

below: On board the *Southern Seas* between the Seychelles and Africa, circa 1950s.

THE ANNALS

AND

MAGAZINE OF NATURAL HISTORY.

[NINTH SERIES.]

No. 56. AUGUST 1922.

XX.—*On a Collection of Mammals from Chiromo and Cholo, Ruo, Nyasaland, made by Mr. Rodney C. Wood, with Field-notes by the Collector.* By P. S. KERSHAW.

(Published by permission of the Trustees of the British Museum.)

THIS interesting collection of beautifully prepared specimens is the result of the labours of some years, and adds very considerably to our knowledge of the distribution of the small mammals of the district. Practically all the collection was made at Chiromo and Cholo in the Shiré Valley, about 17° S., 35° E.

The fine series of Chiroptera call particularly for notice, there being no less than twenty-eight species represented, of which two are new to science. A third novelty is *Uranomys woodi* from Cholo.

Mr. Wood's field-notes are distinguished by inverted commas.

Left:

The first page of the paper by P. S. Kershaw describing Rodney Wood's collection of mammals from Chiromo and Cholo in 1922.

Below:

One example of Wood's extensive correspondence with Dr Oldfield Thomas of the British Museum of Natural History in London; Wood always wrote in longhand and his writing was characteristically clear and precise.

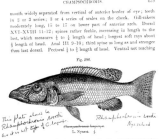

Rodney Wood liked to annotate his books with comments and corrections.

Left: Page 435 of Boulenger's *Fresh-Water Fishes of Africa.*

Middle: Frontispiece of Volume 1 of Boulenger's *Fresh-Water Fishes of Africa* with Rodney Wood's signature. He purchased the four volumes from Boulenger in 1920, and later gave them to Rosemary Lowe-McConnell when she was working on Lake Nyasa, 1945–1947.

Below: Two pages from Belcher's *The Birds of Nyasaland*; Rodney Wood richly annotated more or less every page like this.

Left:
Arthur Westrop, Rodney Wood's friend
for 40 years (Richard Westrop).

Middle:
Magombwa House near Cholo, 1927
(from Westrop's *Green Gold*).

Below:
Rodney Wood (centre in "British Warm") with
members of the Wood Badge Course, Gilwell, May
1923. Arthur Westrop is third from the right in
the back row (from Westrop's *Green Gold*).

Chiromo as Rodney Wood would have known it in 1914 – 1918.

Left:
The railway bridge over the Shire River, used also by motor vehicles and pedestrians (Rees 1910).

Middle:
African Lakes Corporation House (ALC Archives).

Below:
Paddle steamers moored on the bank of the Shire River at Port Herald, south of Chiromo (ALC Archives).

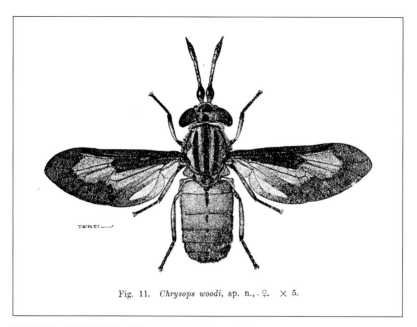

Fig. 11. *Chrysops woodi*, sp. n., ♀. × 5.

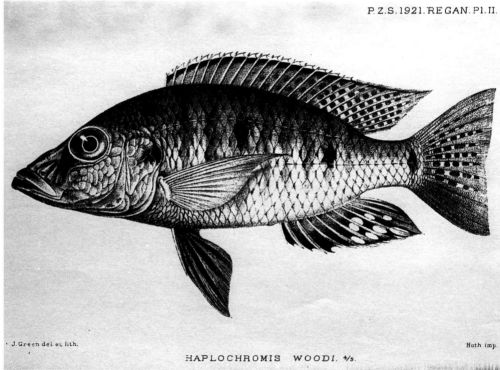

P.Z.S.1921.REGAN.Pl.II.

J.Green del. et lith.

Hath imp.

HAPLOCHROMIS WOODI. ⁴/₅.

Two species named after Rodney Wood.

Above: Chrysops woodi, a tabanid fly (Neave 1915).

Below: Haplochromis woodi, a cichlid fish from Lake Nyasa (Regan 1921).

still moderately abundant in the early years of the twentieth century. Nevertheless, the huge concentrations of animals along the Upper Shire River and on the Phalombe Plain had disappeared.[7] In parts of northern Nyasaland, it was assumed that game was still relatively common, but since this area was less frequented by Europeans, there are few records of species and abundance. In 1932, Murray summed up the situation by stating that 'some years ago big game was very numerous . . . there is now very little game left and what there is very scattered'.[8]

When Wood arrived in Africa, large mammals were still relatively abundant and widespread. With his love of collecting, and his good eye (as evidenced by his ability at playing darts while at school), it is hardly surprising that he became a hunter and enjoyed pitting his wits against those of the animal he was hunting. In 1912, he obtained his own copy of *Central African Game and its Spoor* by Stigand and Lyell, which described all the game animals of central Africa, hunting techniques, and the 'rules' of hunting. As was his usual habit, Wood wrote many annotations in pencil in the book and commented on matters of fact, adding his own viewpoint and making corrections where the facts were incorrect.[9] These annotations, together with other writings, show that he was a very careful hunter – not an indiscriminate shooter – and that his method of hunting relied upon knowing the habits of the animal. For example, he drew lines in the margin next to the following quotations to emphasise their importance: 'There is more to be learned from droppings than from any other department of spooring, and we have deemed it of sufficient importance to collect and endeavour to portray those of most animals'; and ' the sportsman should invariably carry his rifle himself . . .' Some of the annotations illustrate Wood's excellent naturecraft. For example, against the comment 'A collection of spoor of different species had been made [and] although they have been drawn to scale . . . they do not show the resemblance we should have wished', Wood wrote: 'Also I fancy they have muddled up fore and hind feet fairly often.' Wood was always a stickler for using the correct scientific names of animals. On page 154, there is a description of 'Likongwe (Stoat or Weasel)': 'Certainly not,' wrote Wood. 'The Likongwe is the mongoose *Herpestes gracilis*.' He also wrote comments on the distribution of species in Nyasaland; for the cheetah, for example, he wrote: 'Found sparingly over most parts of Angoniland', and for the Gray

Mongoose: 'found frequently by mountain streams in Cholo district and grows to large size'.

There are no detailed records of where Wood hunted, or what species he hunted, when he lived in Rhodesia (1912–1914), nor when he lived at Chiromo and Cholo (1914–1920). He must have obtained a series of good specimens prior to 1920 because he had mounted heads and horns in his house when he was living at Gilwell Park. Some of his specimens must have been very fine because they were listed in Rowland Ward's *Records of Big Game*. These records list the largest horns (in terms of length along the curve, circumference at base, and tip to tip) of each species and give the locality and the name of the owner/hunter. In the ninth edition of 1928,[10] there is a record (undated) of a Nyala from Nyasaland shot by Wood which had horns of 27¾ inches on the front curve, and 17½ inches from tip to tip. In a much later edition, that of 1973, three of Wood's specimens are listed: a Red Lechwe from Zambia, and a Common Reedbuck and a Southern Kudu from Malawi (unfortunately all undated, and without localities). Since there are no dates, one can only surmise that all these specimens were collected sometime prior to 1930.

Wood's hunting methods were rather unorthodox compared with those of the 'typical' hunter. He liked the challenge of stalking and coming as close as possible to the animal before he shot it. He used his knowledge of the species, and his great experience in bushcraft, to creep slowly and quietly through the bush towards his prey. Stalking Nyala was one of his special challenges. In Nyasaland, Nyala only occur in the dense forest and thickets on the western side of the Shire River, not far from where Wood was working on his cotton estate. Nyala are secretive medium-sized antelopes; males are chocolate brown and females are chestnut brown, and both sexes have a pattern of white vertical stripes on the flanks. These colours and patterns make them very difficult to see in the dark dappled light of the forests and thickets where they live. For two years Wood regularly spent his weekends hunting them.[11] When stalking Nyala he took an African gun-bearer with him, and he wore a shirt made up of patches of different shapes and colours, with twigs fastened to his hat and over one shoulder to break up his outline. Such attention to camouflage is not unusual now, but very few hunters took such precautions in those days. Stealth, quietness and perseverance were essential while looking for Nyala. Wood's own words convey his enjoyment and happiness while hunting them:

> Blue Duiker and Livingstone's Suni were often seen, and more often heard, giving their curious nasal 'Nsss' cry of alarm as they scampered off in the undergrowth, as my native gun-bearer and I crawled on our knees or wormed our way on our stomachs along the tunnels into the jungle made by Bush Pig and other game. After the first few yards through the tangle of thorny scrub and creepers, we were in a land of semi-darkness where grew little ground vegetation, but only the tangled and intertwining stems of many climbers and lianas. Monkey-ropes canopied the trees overhead with a dense blanket through which the sun could hardly penetrate. The forest giants broke though this canopy and spread their massive limbs of foliage over wide areas, adding to the gloom. The whole place reeked of moist heat and decay, and often the very air was stifling. Birds and monkeys used the main canopy as an aerially suspended world of their own, and the forest resounded with their weird cries and yells. Occasionally one would hear the deep roaring bark of a startled Nyala bull – Bo-o-o-o-gh – from which sound their local name 'Mbo' is derived with onomatopoeic allusion. An eerie place, but fascinating with an allurement all its own. One came upon little glades here and there where a few beams of sunshine gave cause to some flowers and grass to struggle for existence, and these were favourite spots for the Nyala. One moved as a shadow in this land of shadows, and an occasional twig-snap showed where the Nyala bull faded as a shadow among the tree stems.[12]

Wood waited a long time before he was satisfied that he had the perfect shot. On thirty-two previous occasions he had seen Nyala in the forest, but none of these offered the certainty of the shot that drops the animal where it stands. Then, on the thirty-third occasion, after careful observation and a lot of experience, Wood found a Nyala at dawn where he expected it to be. He had no sympathy for the hunter who took 'the chance shot', and he considered that such a shot is 'criminal at all times' and especially when the animal is rare and little known.[13] The story of the shooting of this particular Nyala is one of the most remarkable tales of game hunting in Africa, and it also tells a lot about Wood's attitude to hunting:

Then, after two years of endeavour and patience, came the knowledge of that link in the chain of their lives which spelt success for the hunter.

Since before dawn we had crept along our usual track, every inch of which was now known so well, and were now, at about eight o'clock, with the sun well up, slowly walking back towards camp through the forest, a short distance from the jungle wall. The fires had been through some weeks previously, though only a poor burn had resulted as they were too early to make a clean sweep. But now several species of trees and shrubs were in flower, giving a sweet perfume to the hot air.

In a somewhat dense piece of forest stood a large tree with long racemes of purplish-red flowers. Around it was a small open space with comparatively bare ground beneath the tree itself. There had been a slight shower of rain the day before and automatically, without any special thought, I moved to the spot to look at whatever spoor might be marked there. To my surprise, the ground beneath the tree was covered with the tracks of Nyala, both bulls and cows, and those tracks were fresh, some showing that a bull Nyala had only just dashed off at our approach. This was confirmed by droppings that were yet quite moist and warm.

For a few moments, I stood looking around, not yet fully comprehending. What on earth were Nyala bulls doing under this tree well after sunrise at somewhere about eight or eight-thirty in the morning? And relatively so far away from the main jungle mass? . . .

As I stood endeavouring to unravel it, a large flower-petal dropped to the ground from the tree. Something was holding me, subconsciously worrying my brain since my arrival on the spot, but the concrete idea was elusive. Another flower fell – a few seconds later a third. In a flash I understood . . . I knew the tree well with its habit of shedding its flower-petals one at a time every few minutes when in full bloom. They were large and sweet-smelling, and I had often noticed native cattle and goats in the villages eating them greedily. Three had fallen while I stood there, but there were no more lying on the ground! So the Nyala had been eating them as they fell and, what was more to the point, the attraction was so great that they threw away their usual caution and stayed by the tree, long after daylight!

I returned to camp, knowing that the long quest was drawing to its end at last. Long before dawn the next morning, I was concealed in a good position among some long grass and scrubby undergrowth nearby, having crept up quietly along an approach I had well reconnoitred the day before. Between me and the tree with the flowers, was a stretch of broken and beaten-down grass, hiding from view the bare patch below the tree, but not high enough to entirely conceal the body of a Nyala bull. I carried a double-barrelled .475, as I meant to make sure of a clean kill. Dawn came, but I must wait till daylight before rising without a sound to peer towards the tree. Those minutes were hours but I kept myself in hand. At last I could see objects clearly enough. The great moment had come; broadside on below the tree was a fine bull. He fell at a shot through the neck, dropped out of sight behind the grass, but recovered again at once and dashed past me, dropping dead to the left barrel only a few yards from me.

The gods of hunting, that had tested my patience for so long, were indeed to glut me that fateful morning. Examining the dead bull, it seemed to me that the horns were somewhat shorter than I had thought on first seeing it. Going up to the tree I found on the bare patch a bigger bull lying stone dead. A few yards beyond him lay a third, killed by the same bullet. The one killed with my second bullet had been another animal altogether which must have been lying down at the feet of the first one and had sprung up at the first shot, dashing past me in panic and so making me think it was the same animal. The bull killed with the same bullet as the first one must have been in the line of the shot, of course quite invisible to me behind the other. Three Nyala bulls dead with a right and a left – it seemed to me horrible. Never before in years of almost daily hunting, had I ever bagged two animals with one bullet, and now for this – and worse – to happen with Nyala was unthinkable. All my joy at the longed-for success had vanished.[14]

This passage was written many years after the event, but it was a turning point in Wood's life – he never hunted Nyala again. The sequence of events which led to the killing of these bulls confirmed his belief that it is easy to kill a rare and shy animal once its habits are known. As a result of this remarkable incident, Wood worked for the conservation of the Nyala and its habitat. In due course, the government made the hunting of Nyala in Nyasaland lawful only with a special licence, and, in 1928, proclaimed two game reserves which protected the main habitats where they breed. So perhaps, as Wood said, 'those three glorious bulls did not die in vain'.

One of the major topics of conversation and debate in Nyasaland for many years has been the relationship between tsetse flies, wild animals and sleeping sickness. Tsetse flies are widespread in tropical Africa, especially in the savanna zones. When he first went to British Central Africa, Sir Harry Johnson, thought that tsetse flies were large, colourful, ferocious flies with a long biting proboscis.[15] In fact, they are brown in colour, and smaller than the large horse flies which are also found in savanna habitats. Johnston recorded that tsetse flies could give a nasty painful bite and that cattle died if 'bitten three or more times'. As far as humans are concerned, he wrote: 'Its bite on man produces absolutely no effect beyond the pain of a sharp puncture.' Johnston's ideas about tsetse flies were, of course, quite incorrect, but his thoughts make interesting reading:

> The nature of the tsetse poison is not yet determined . . . some advance the theory that there is no inherent poison in the tsetse itself, but that it inserts the germs of malaria. These, passed on by the tsetse fly, passing with infected proboscis from wild to tame animal, increase and multiply in the latter, which is not inoculated [sic], and the beast dies not from a specific 'tsetse' poison but from malaria introduced by the tsetse. I confess, however, this theory, although ingenious, does not strike me as adequately accounting for all the facts. I can not help thinking myself that the tsetse must secrete and introduce into the animal's system a peculiar venom which in the human being causes the bite to itch. . .

However, soon after Johnston had written this, research in other parts of Africa showed that tsetse flies are responsible for transmitting microscopic single-celled organisms called trypanosomes from one host to another host. Trypanosomes are the cause of the condition trypanosomiasis, commonly called 'sleeping sickness' in humans and 'nagama' in domestic cattle, horses and goats. There are several species and subspecies of trypanosomes, of which *Trypanosoma vivax* and *T. brucei* are the most important and widespread. Wild mammals do not suffer any pathological condition when parasitised by trypanosomes, but they act as 'reservoir hosts', providing a constant pool of trypanosomes which could infect humans and domestic animals. When a tsetse fly bites an infected wild mammal in order to obtain a blood meal, it takes up trypanosomes through its proboscis. Later, when the tsetse takes its next blood meal, some trypanosomes pass from the tsetse into the blood of the next host. The trypanosomes reproduce and multiply in the blood of the host, and if the host happens to be a human or domestic animal, trypanosomiasis will develop. The cycle of host to tsetse fly to host is continuous, but the incidence and frequency of trypanosomiasis depends on many variables, such as the number of reservoir hosts, the number of tsetse flies, the climate, the season of the year, and whether there is suitable habitat for tsetse flies and game animals. In Nyasaland (and Africa as a whole), the distribution of tsetse flies is patchy because adult tsetse flies, and their pupae, require very specific conditions of temperature and humidity to survive. Some areas of Nyasaland have tsetse flies, notably the Lower Shire Valley south of the Elephant Marsh, around Liwonde and Fort Johnston, on the plateau around Dedza, Lilongwe, Dowa and Fort Manning, and along part of the western shore of Lake Nyasa. Other parts are fly-free, including all land above about 1,000 metres. Johnston observed, correctly, that tsetse flies do not occur over water and do not fly at night; and it was with some pleasure that he wrote that 'it is possible to convey horses and cattle up the rivers in midstream without the least danger of them being bitten' and that if 'a tsetse-haunted district must be crossed, it should be done at night-time – by moonlight if possible'.[16]

Once the lifecycle of trypanosomes was understood in the early years of the twentieth century, many administrators, doctors, and missionaries considered it was essential that game animals should be eliminated for the good of humanity. Their argument was that if game animals were destroyed, tsetse flies would disappear and sleeping sickness could no longer be transmitted to humans and domestic animals. At times, in other parts of Africa, sleeping sickness had reached epidemic proportions – in Uganda and the Congo, for example – and the government was concerned that the spread of tsetse flies in Nyasaland would increase the prevalence of the disease. Other measures that were introduced to combat the disease before the First World War included taking blood smears from travellers entering from the north, isolating people with sleeping sickness, clearing bushes around villages and along roads, and building proper latrines in clearings.[17] The view that elimination of game was essential for tsetse control is best summed up in an editorial in the *Nyasaland Times* on 2 January 1913:

> . . . it seems to be a proved fact that the Common Tsetse Fly which was long suspect is now convicted beyond doubt as being the main carrier of sleeping sickness. It is also proved evidently that the main hosts are big game. . . . We have been repeatedly assured that, on proof being forthcoming that Tsetse and Big Game being the culprits, Government would have no hesitation in taking the most stringent measures to rid the country of this menace to the lives of the people of this Protectorate. . . . We have laboured this subject so much before . . . all we wish to say now is that the community have a right to demand the most stringent measures to get rid of both game and tsetse. Once again, we repeat the easiest way to get rid of the Big Game is to give carte blanche to shoot every variety outside the game reserve.[18]

And on 18 August 1913, after pointing out that there had been several Commissions which had shown the relationship between game, tsetse flies and sleeping sickness, another editorial continued:

> To the ordinary layman, the question has always presented itself as a simple one Game versus Humanity. . . . As we have said so often, we have no antipathy to Game *per se* in their proper place. By all means keep a reserve and fence them in as is done in other countries . . . Remove all restrictions on hunting and the Reserve will soon be full of those which survive and the [Africans] will make short work of those which are outside the limits. It was another excellent piece of fooling which said it was impossible to clear a given area by drives of Askaris. No doubt it was, but that was showing how NOT to do it. The only sure and certain method – and the Government will not take it simply because it is the only practical method – is to remove all restrictions outside the Reserve and let the Game go. . . . there is a powerful and influential group which opposes any talk of extermination and which evidently exhaust every possible subterfuge before giving in. . . . Doubtless we will be told a Commission will have to be held to consider the point (i.e. that natives as well as Europeans must be allowed to 'hunt, harry, net

and kill' as many game as possible), and doubtless, too, future generations will think a Commission should have been held to consider the
sanity of this present government.[19]

The government did not sanction the wholesale slaughter of big
game, but maintained the Game Regulations of 1902, 1906 and 1911
which required a hunter to obtain a licence which allowed him to shoot
a specified number of individual animals of listed species. The licence
had the effect of restricting the hunting of game to Europeans, and
was not particularly concerned with the protection or conservation of
species.[20] At this time in history, no one imagined that any species of
large African mammal could become endangered or would need 'protection', nor that the face of Africa would change so much in so short a
time.

After the First World War, the future of game and tsetse flies was
hotly debated again when the government was in the process of revising
the Game Regulations. Nyasaland was changing rapidly – new estates
were developing, the African and European populations were increasing,
and the economic prosperity of the country was improving. Undoubtedly
wild mammals were a nuisance in certain situations; some species, such
as elephants, baboons, and small antelopes, were causing losses to agriculture, while others, such as lions and leopards, were dangerous to
humans, and many species were considered to be reservoir hosts to
organisms that caused disease in humans and domestic animals. For these
reasons, there was a vocal section of the community which believed (as
it did in the years before the First World War) that most, if not all, game
animals should be destroyed and that the Game Act, which gave protection to animals, should be repealed. The missionaries considered that
any legislation should concentrate on the removal of game rather than
on its protection. They based their case on four main arguments: (a) wild
animals cause damage to crops and since local Africans are banned from
owning firearms, they are unable to protect their crops adequately, (b)
large predators are a danger to human life, (c) the Act forbids traditional
means of hunting and ownership of guns, and therefore Africans are
deprived of their traditional source of protein, and (d) game animals
harbour diseases which are transmitted to humans and domestic animals
by tsetse flies.[21] Rather surprisingly, many of the planters joined the

missionaries in their condemnation of the proposed regulations because they required that a licence was needed to kill an animal on any land, whether owned by the government, an African or a European. Planters regarded any animals living on their estates as their own property, and did not take kindly to the idea that they had to pay for a licence to shoot on their own land. Many of these arguments were taken up by the *Nyasaland Times*, since the editor of the time was also vehemently opposed to the proposed regulations. In the editorial of 2 March 1926, he wrote:

> The government has decided that the new Game Ordinance shall come into operation on April 1st next . . . It is a little difficult to conjecture how the government hopes to benefit by the enactment of such a measure. If it is expected that an increased revenue will accrue, we venture to suggest that it is hardly likely to succeed; if it is intended solely as a measure for preserving the fauna of the country, it is a distinct disservice to both [African] and European, more particularly to the former, perhaps, for [he] is always the main sufferer from the depredations of wild animals because he is unable to protect his crops in the same way as the European. It is a recognised fact that wild animals and tsetse fly are concomitant, and it is also true that large tracts of this country have been depopulated by this pest – good land that has been devastated and made impossible of habitation for both man and beast. And yet the government persists in preserving big game when they well know that this very policy has in the past led to the re-introduction over large areas of the tsetse which had disappeared as a result of the rinderpest epidemic. *This spreading of the tsetse is still going on and no half measures will eradicate it.* There is only one practical method, and that is to allow free hunting and shooting of all big game by [African] and European alike [editor's italics].[22]

Wood contributed to this debate by writing several letters to the *Nyasaland Times*. He, as might be expected, took the opposite view. On 2 March 1926, he wrote:

> Sir – All those interested in the conservation of wild game animals of this country will welcome this new Ordinance, upon which the Government are to be congratulated. To those who remember the abun-

dance of our large game before the war, and who have remarked on the appalling rate of its diminution in recent years, it has been obvious for some time that the existing Ordinance had become obsolete as affording protection, and that drastic new regulations were necessary if the game were to be conserved.

There will doubtless be discussion of the Bill among people of this country, so it might be of general interest to point out a few of the principles of conservation which many international Game Conservation Commissions have postulated as basic for the preservation of all species, upon which principles this Bill is undoubtedly based . . . It is an accepted fact that among all enlightened nations that the wild game of any country is held by the people *in trust*. It is not theirs to destroy wantonly. They are accountable for it to their children and posterity. Its value – aesthetic, economic, creative, fitness and efficiency promoting – is admitted by all without being open to question, and our children have an equal right to it. It has been found throughout the world that proper and effective conservation must be based on the following unalterable fundamental conditions:

(a) Protection given in time, before any species has become dangerously near disappearance. No species must be destroyed at a greater rate than it can increase.

(b) Absolute protection of all females.

(c) Adequate reserves where no form of molestation whatsoever may take place. Local sanctuaries wherever possible.

(d) Protection from unwarranted slaughter, i.e. provident utilisation as against injudicious exploitation.

(e) Legislation to prevent the killing of too many *virile* males, i.e. regulations of the number that may be killed.

> Points that should always be borne in mind are that local abundance should never be taken as an indication of general abundance, or as a reason for permitting killing in large numbers, as animals are apt to bunch at certain seasons or if unduly molested. Secondly, when animals become reduced in numbers through the onslaughts of man, then their natural enemies, not similarly checked, take abnormal toll. Thus do species become suddenly extinct [Wood's italics].[23]

This is a remarkable letter because it is far in advance of its time in terms of the conservation values it is founded on. Wood's knowledge and his keen advocacy for conservation in central Africa show that he had a good grasp of scientific principles and was not, like his protagonists, swayed by emotion. However, he did not consider the vexed question of the relationship between game, tsetse flies and disease. The correspondence continued, almost on a daily basis, with most contributors recommending destruction of game, since this would be followed, according to them, by the disappearance of tsetse. Wood could not allow such statements to go unanswered, so on 12 March 1926, he wrote another letter to the *Nyasaland Times*:

> Sir, Permit me to reply to some of the remarks in your editorial article and other letters on this subject. I have no intention of entering into argument on the vexed subject of the relation between game and tsetse fly, but would like to take this opportunity of pointing out a few facts. . . . Although all observers and investigators agree that *Glossina morsitans*, or tsetse fly, prefers game to any other food when it can get it, *there is no definite proof whatsoever* that the fly can not exist or would disappear completely if the game were exterminated. It has often been shown that it then becomes much more ready to attack man for food and also feeds on small mammals such as squirrels, rats, bats etc., as well as baboons and monkeys and sometimes birds. It would be interesting to know how the opponents of the Bill, on 'fly' grounds, intend to exterminate completely all these small mammals, as it has been shown by many investigators that wherever any mammalian blood persists so will the fly. One thing only is yet proved conclusively by all which is that where the forest haunts of the fly are destroyed, so also are the fly. . . .
>
> With regard to the diminution of the fly after the rinderpest epidemics [of 1892–1896], I have before me a mass of evidence that *this only took place in some localities and not in others*. Our own former Governor, Sir Alfred Sharpe, a close observer and keen hunter, writes that when the rinderpest visited Nyasaland there was a noticeable decrease in the numbers of big game in some of the fly-belts but not in others, especially of buffalo, but goes on to say 'but I have never noticed that this had any effect on the Glossina'. Stevenson-Hamilton remarks, 'while there is no doubt that G. morsitans (?) absolutely disappeared from considerable

areas during or immediately after the epidemics, we have it on reliable authority that *elsewhere it was in no way affected.*' And so on with other investigators.

It has now been generally suggested that it was an effect of the dry season (in many places a drought) coupled with the very sudden cutting off of its food supply through the dreadful mortality of game that affected the breeding of the fly in certain localities, and this combination of circumstances caused their decrease. Where there was no such combination, the fly managed to remain as usual.

I point out these facts because it evident how very one-sided this argument can become unless *all evidence* is stated. . . .

I utter this solemn warning. That if the government and people of this country permit the destruction of their game resources, and it is then found that no good whatsoever had been done towards the extermination of the 'fly', then they will have been guilty of the greatest mistake and disservice (I almost wrote 'crime') to the future inhabitants of this country and posterity. Once gone, the great game fauna can never be replaced . . . [Wood's italics].[24]

In the end, the government of the day stuck to its guns and introduced the 1926 Game Regulations. One suspects that Wood had more to do with the new Regulations than is suggested by his letters. He was well known for his views on conservation and had been involved in conservation issues in the past; in 1921, for example, he was chairman of a committee which investigated the indiscriminate shooting of game for meat, and the source of large amounts of game meat on sale in local markets, especially in Blantyre and Zomba.[25] And in the late 1920s, when he was the Honorary Game Warden, he was consulted by the Governor about trout fishing on Zomba plateau.[26] No doubt his friendship with the Governor and the Chief Secretary enabled him to put his views to those who made the laws of the country.

Looking back on these heated discussions of the 1920s, one can see that both viewpoints had their merits. However, much more is known about tsetse flies now than in those days. It is now known that tsetse flies do not feed on all species of game animals as was supposed. Analysis of the blood in the stomachs of tsetse flies in one area of East Africa showed that 77% of meals were from warthogs and 14%

from buffalo. Tsetse flies rarely fed on many other species of mammal which were also present, such as impala, waterbuck, did-dik and harte-beest. Moreover, although impala formed about 70% of all mammals, they formed only 1% of the blood meals of the tsetse flies.[27] Each species of tsetse fly has slightly different preferences from the others, and these vary according to the locality. In general, tsetse flies feed pref-erentially on various species of pigs, bushbuck and buffalo, and tend not to feed on duikers, dik-diks, waterbuck, gazelles, impala, hartebeests, or baboons.[28] Although there is no precise information for Nyasaland, the principle is the same: tsetse flies are selective, and hence the whole-sale killing of game as a means of controlling tsetse flies is pointless. Nevertheless, thousands of game animals were killed in tsetse eradica-tion schemes in various parts of Africa in the 1930s, 1940s and 1950s, and none of these schemes succeeded in eliminating tsetse flies. There was some reduction of tsetse flies to begin with, but they quickly recolonised from other areas. These eradication schemes also showed that many species of mammal – especially the small and fecund species – can increase their numbers very quickly after their population has been reduced. Over the years, and up to the present time, it has been shown that the best methods to reduce the problems of tsetse flies have been clearing habitats where flies live, spraying chemicals in fly-infested habitats close to human habitations, creating buffer zones between fly-infested areas and settled areas, and reducing the chances of contact between flies, humans and domestic animals. None of these methods can be done on a large scale – it is quite impossible to eliminate tsetse flies from Africa. If the 1926 Regulations had been repealed, and shooting of all game animals throughout the country had been allowed, it is likely that there would be no game animals (or extremely few) in modern Malawi.

Wood was appointed as Game Warden in Nyasaland on 1 May 1929.[29] Although this term was generally used by most people, the official desig-nation of the appointment was 'Game, Cultivation Protection, and Tsetse Fly Control Officer'. He was the first Game Warden to be appointed in Nyasaland, so part of his job was to set up a Game Department. He also had to do a lot of fieldwork, investigate tsetse fly distribution, prepare blood slides, and organise teams of labourers to clear bush. All this involved a lot of travelling, and for many months at a time he was constantly on the move from place to place.[30]

Most of Wood's work on tsetse flies was on the Central African Plateau in the centre of Nyasaland. Here, the Bua River and its tributaries flow eastwards across the plateau before descending to the lake. The country is rather flat and covered with savanna woodland; small drainage channels often fan out to form extensive swamps or *dambos*. In places, little rocky isolated hills rise up above the plateau; from the tops of these hills, there are marvellous views over hundreds of square miles of woodlands, swamps and rocky hills. This is perfect country for tsetse flies because it has lots of cover as well as moisture; it is also good for game animals because of the abundant grass, shrubs and water. Likewise, it is good country for cattle and for agriculture.

Wood set up his first headquarters in the Assistant District Commissioner's house at Dowa, although this soon proved to be inconvenient and he moved to Lilongwe. For the first six months, during the dry season, he spent much of his time on *ulendo*, or safari, along the Bua River, setting up camps from where his many assistants could carry out fly surveys. During these surveys, they walked along the small paths that linked the villages (each village being named after its chief) and along the edges of the *dambos*. Wood's diaries[31] from these days give a good idea of the sort of life he led on *ulendo*:

24.vi.29. Left Dowa on ulendo about 10 am. Tiffin by Lingadzi to west of Kongwe Mountain on Ngara old road. Camped at Mdajola's some 10–12 miles out, where much beer and a chinamwai dance at a nearby village, the whole night being an uproar of drumming, singing and shouting. Many cattle all round this district of the Kabinde stream.

25.vi.29. Up early and on to Musa's village about another 12 miles where camped as the carriers appeared very done in and footsore, and they said they got nothing to eat last night. Only well-water here. Forest of largish trees, good soil and long chipita grass. No cattle, as fly said to be numerous here. Saw spoor of sable just north of Dwole's.

26.vi.29. From Musa's through Chimungu's on Chawawa dambo northwest to Chimkwiri's, some 5 to 6 miles. Only a deep well in dambo here. Tsetse first noticed in village in dambo. At 2 pm out for game towards Mpali Mountain, but grass not burnt except in one dambo where masses of old elephant, buffalo, roan and other spoor, and fresh tracks of a rhino. Saw three zebra together, a male, female and very young one. Shot a duiker. . . . Back at dark to camp. Dreadfully bitten on legs by 'fly' all afternoon which very numerous, tho' little game seen. Scores of guinea fowl on burnt dambo.

27.vi 29.   Chimkwiri's to Kasense's on Kasangadgi Road. Fly all the way, which through useless 'msanga' country of worst type. Fair amount of game spoor. Came on herd of eland in open dambo – only cows and 3/4 grown young. Easy stalk so shot two for meat supply for the ulendo . . . felt like a murderer in doing it, but carriers must have meat when possible. Also shot a warthog for blood slide examination for tryps. My legs so swelled up with 'fly' bites in calves that almost crippled for last three miles . . . old tracks of elephants everywhere . . . No flies were by the eland, and none seen by the two dead ones, nor on the warthog. This was noticeable as fly were with us all the time on the march.

Wood spent a day or two near Ngara mountain, close to where the main road (at this time just a dirt track) going north crossed the Bua River. Here he commented that 'the early hours of the morning were hideous with lions in all directions, and I had to get up to try to locate one at the back of the camp and whose stomach rumbled'. From Ngara, he continued westwards upstream along the Bua River, spending the best part of two weeks walking from village to village, noting the presence or absence of tsetse flies, and taking blood smears from animals he shot. He took samples from duiker, eland, puku, hartebeest, waterbuck, oribi, reedbuck and warthog. He was impressed by the amount of game, and commented that often there was plenty of game and no fly.

On another *ulendo*, in July–September 1929, Wood walked back along the Bua River in order to set up a permanent camp where labourers could live while cutting fly lines through the bush. He selected a campsite at the base of Msula Hill, about one mile from the Bua River, for his headquarters in this area; the camp was given the name of 'Elephant Butte' because of all the elephants seen in the area. First a track had to be built so that cars could be driven from the nearest road; huts had to be built and local villagers enlisted as labourers. Much of Wood's time was spent on the administration of the camp, purchasing and allocating *posho* rations, keeping the labour books, writing official correspondence, shooting game to provide meat for the labourers, organising the cutting of fly-lines, placing baits to attract lions, and attending what he called 'pow-wows' with village chiefs. To Wood, no doubt, cutting lines in the bush and trying to keep tsetse flies away from humans and cattle was preferable to the wholesale slaughter of game, which the opponents of the 1926 Game Regulations would have liked.

The amount of game in the Bua catchment was impressive, and Wood often recorded what he saw on his travels. In his diary for August 1929, he wrote:

> Two lions passed along road by camp in night; no-one heard anything . . . A rhino grazed and watered opposite camp at night . . . Scores of puku, oribi, reedbuck, waterbuck and hartebeest . . . Large herds of 18 waterbuck and 11 zebra together . . . Alive with Puku everywhere; literally hundreds; very tame and foolish, taking little notice of wind . . . Camped in lovely circle of big trees and thickets with Puku and Waterbuck grazing all around – former in scores and one herd over 60.

During the years that followed, most of these large herds of game animals disappeared. Wood probably realised that this would be an inevitable consequence of settlement and agriculture, and hence he made plans for a relatively uninhabited part of the Central African Plateau within Nyasaland to be set aside as a protected area. In February 1930, he marked out the boundaries of a game reserve[32]. This was gazetted as Kasungu Game Reserve later that year, and in 1970 it gained the status of a National Park.

In addition to his work in the Bua catchment, there were other projects which occupied Wood's time. To prevent tsetse flies invading the Plateau from Portuguese East Africa, a wide fly-control belt was established along the road which runs westwards from Lilongwe to Fort Manning close to the border. At one stage, he had about 200 labourers clearing the bush on the north side of the road, and he was pleased to note that no flies were seen on the road after clearing and that there was no evidence of game crossing the road where vegetation had been removed. On other occasions, Wood was involved in plans for the creation of a 'sanctuary' on the islands and mainland around Monkey Bay at the southern end of Lake Nyasa, and for the modification of the Mulungusi stream on Zomba Plateau to make it more suitable for trout (as well as for trout fishermen!).

Wood soon became aware that having a base at Dowa was unsatisfactory, so he moved to Lilongwe, then just a small town. Since there was no suitable accommodation, he built himself a 'shack' by the Lilongwe River which served as both home and office. Many days were spent on 'office work, building, etc', as well as checking on the progress of the fly gangs. However, even Lilongwe did not meet his needs, and five months later he moved to Zomba, the government headquarters, where he was close to all the officials who could sanction his plans for the Game Department. One of the advantages of being Game Warden was that it gave him the opportunity to meet many interesting people. He was a good friend of the Governor, Sir Shenton Thomas, and frequently stayed at Government House; he worked closely with Dr W. A. Lamborn, the Government Entomologist who lived at Fort Johnston, and with C. H. Wade, the Assistant Chief Secretary. Because of his position, he gave talks to various groups and colleges on game conservation, mammals and birds, and trees, and he also wrote the chapter 'Animal and Bird Life' for the 1932 edition of the *Handbook of Nyasaland*.[33] He was evidently well known in hunting circles outside Nyasaland, because Major Mayden asked him to contribute two chapters on Nyasaland for his book *Big Game Shooting in Africa*, sections of which are quoted above. Other contributors to the book included many well-known hunters, such as Lieutenant-Colonel R. E. Drake-Brockman, Denis Lyell, Major P. H. G. Powell-Cotton, Colonel Stevenson Hamilton, Captain G. Blaine and Captain A. T. A. Ritchie.

Wood did not manage any serious scientific collecting while he was Game Warden, although he did collect a few butterflies. All of his collections of birds, butterflies and other insects (other than those already sent to the British Museum of Natural History) stayed at Magombwa while he held this position. In September 1929, while Wood was camping at Elephant Butte, an American couple, Mr and Mrs Rudyerd Boulton, turned up at the camp. Boulton was a collector from the Carnegie

Museum of Natural History in Pittsburgh and was collecting birds from eastern Africa for the museum. Wood must have been delighted to meet a fellow collector and he described the Boultons as 'a delightful couple'. The Boultons camped with Wood and they had 'a splendid evening together', and later Wood left them at Elephant Butte so they could collect birds along the Rusa River. This chance meeting had important implications for Wood's collections, because Boulton persuaded Wood to sell most of his bird specimens to the Carnegie Museum (see Chapter 11). Later on, Boulton visited Magombwa, where he and Wood packed the specimens ready for their journey to America. It is surprising that Wood agreed to sell, but at this time (and, indeed, throughout most of his life) he was slightly short of money. Also, he knew that he was leaving Magombwa and going, eventually, to the Seychelles, and he would have known that the specimens were going to a museum where they would be well curated and available to researchers on African animals.

Wood was able to accomplish the work he did because he was extremely well organised. He built up a network of friends with whom he could stay; such 'dropping in' was a standard way of African life in those days, and the sudden arrival of a guest for a night or two was not seen as an imposition because everyone had house servants and gardeners. Travelling was relatively quick and easy because, in the years since the end of the war, the road system had been improved and extended – even though most roads were still made of dirt, and rivers and stream were crossed by drifts rather than by bridges. In the wet season, many of the roads flooded and some were impassable. Wood had two vehicles, a Morris Cowley and a Douglas truck,[34] and since he held a driving licence, he probably drove them himself. But there were always problems with them, and on several occasions Wood had to replace the springs of the vehicles – a telling comment on the state of the roads! There were many other mechanical breakdowns, such as punctures and leaking radiators. When travelling, Wood took his firearms – three rifles, a shotgun, and three revolvers – for collecting animals for blood samples and meat.[35]

Life as a Game Warden involved a mixture of office work and field work. After spending the whole of September and early October 1930 in Zomba doing office work, it must have been a relief for Wood to 'escape to the bush' for an *ulendo* with the Governor:

> 10.x.30. To Dedza by Government car, but owing to breakdown only got to Ncheu. Stayed night with Murphys.
>
> 11.x.30. On to Lilongwe. Leaking radiator – bad journey. Stayed night with Vassall.
>
> 12.x.30. To Mudi where night with Lewins.
>
> 13.x. 30. To Governor's camp at Kalambwi, near Mzama.
>
> 14.x.30 and 15.x.30. Attending to erection of camp, getting in some meat (1 Reedbuck male, 1 Waterbuck male), etc. Thunderstorms. Governor and Mrs Thomas arrived 11.30 am. Out with H.E. and Mrs T. after tea along dambos at north of Luntwi; puku and water-buck.
>
> 16.x.30 to 20.x.30. Kalambwi camp on Bua. With H.E. and Mrs Thomas all the time. Saw much game around, chiefly Puku, Waterbuck, Kudu, Reedbuck, Oribi, Warthog, Buffalo. Spurwing, Knobnose, Pinkbill, Yellowbill and Hottentot Teal very tame on river pond by camp all day long, taking no notice of us. Lion around one night; leopard around one night. On 20th we struck camp and returned to Lilongwe.

Wood, together with the Governor and Mrs Thomas, then returned to Zomba. Such trips, and the many meetings that Wood had with various Governors both before and while he was Game Warden, allowed him to present his views on animal conservation in Nyasaland to the highest authorities in the land. He was also a friend of C. F. Belcher, Chief Justice of Nyasaland and a very ardent ornithologist; as Chief Justice, Belcher was responsible for the drafting of legislation concerning game conservation. These contacts undoubtedly influenced the course of events because many, if not most, of Wood's ideas were accepted by the government and passed into legislation. With so many successes, it must have come as a surprise to the government when Wood tendered his resignation in early December 1930:

Sir, I have the honour to request permission to resign my post as Game Warden. I take this step solely because I find the condition of my health does not permit me to carry out the duties of Game Warden in the manner which I feel they should be performed. For the past year I have felt very badly the strain of travelling in all parts of the Protectorate under conditions which must of necessity be at times very rough and hard, and in consequence my health has suffered considerably. Being now over 40 and having had 21 years of tropical Africa is probably the explanation. . . .[36]

One wonders whether the dullness of office work, the bureaucracy of government, the lack of opportunity to collect, and his desire to return to the Seychelles also influenced Wood's decision. The Chief Secretary minuted Wood's file: ' Mr Wood has discussed this with me and has quite made up his mind to go . . . Mr Wood has interesting suggestions for continuing economically the important features of his work.' When the file reached the Governor, Thomas wrote (in his usual neat handwriting and in red ink): 'I am very sorry – we shall all be sorry – to learn of Mr Wood's desire to resign, and its cause. He has served the government well during his short tenure of his office. . . . I should like to discuss the future with him next week.'[37]

Although Wood spent only twenty months as Game Warden, he made many notable contributions which have had a long-term effect on game conservation in Nyasaland. He set up the embryonic Game Department which, in future years, became the Department of National Parks and Wildlife; he organised bush clearing as a means of controlling the spread of tsetse flies; he collected blood samples from game animals, which assisted in assessing the role of game animals as hosts of trypanosomes; and he was a catalyst for the legislation of the Game Regulations of 1927 which disallowed the wholesale shooting of game. Three of Malaŵi's current National Parks were established as conservation areas during (or just after) the time when he was very active in influencing public opinion and when he was Game Warden. Lengwe Game Reserve, known for its magnificent herds of Nyala, was established in 1928 and became a National Park in 1970;[38] Kasungu Game Reserve was established in 1930 to protect the fauna of the Plateau and became a National Park in 1970; and the islands and mainland around Monkey Bay and Cape

Maclear were gazetted as Forest Reserves in 1934 and became Lake Malaŵi National Park in 1980.

Wood left Nyasaland on 27 January 1931 by train for Beira on the Indian Ocean coast. He travelled the well-worn route of all Europeans when leaving the Protectorate: he boarded the Shire Highlands Railway at Blantyre and travelled south through Luchenza (not far from Magombwa), down the escarpment to Chiromo (past his old home on the banks of the Ruo) and along the east bank of the Shire River to Port Herald. Here the train continued along the tracks of the Central African Railway, first opened in 1915, for the 61 miles to Chindio on the north bank of the Zambesi. This part of the journey took all day, from 7 a.m. to 5.30 p.m. At Chindio, passengers embarked on the ferry for the overnight crossing to Marraca. Dinner and a bed were provided for each passenger on the ferry. The final stage of the journey was on the Trans-Zambesia railway from Murraca to Beira, which took another full day, from 7 a.m. to 6.25 p.m.[39] Two days and one night for the journey from Blantyre to the coast was a far cry from the weeks that it had taken 25 years previously. At Beira, Wood boarded the SS *Karapara* for the sea journey to the Seychelles via Durban.[40] This was the first of many visits to the Seychelles, where he lived – on and off – for some nine years, between the 1920s and the 1960s. His life in the Seychelles, so very different from that in Africa, is recounted in a separate chapter.

He did not return to Africa until October 1932.

Chapter 7

AFRICAN WANDERINGS: 1932–1942

Wood left the Seychelles on 13 September 1932 on the SS *Karanga*, one of the British India Steamship Company's boats which plied between Africa and India. There were only ten passengers, and Wood had a large cabin to himself. Travel on ocean liners in those days was pleasant and leisurely; there was good company and good food, and the chance to go ashore whenever the ship was in dock. For most of the time on this particular trip the sea was calm, but at one stage there was a heavy swell and all the passengers except Wood stayed in their cabins; he evidently enjoyed this, and recorded the fact that 'at dinner I was the only person in the saloon at all; first time I've been in solitary state on a 10,000 ton liner!' After stops at Mombasa, Zanzibar and Dar es Salaam, the *Karanja* arrived in Beira on 24 September. Here, Wood boarded the train for the two-day journey to Bulawayo, which he did not enjoy. In his diary, he wrote: 'Journey very hot and dusty by day and foully cold at night; v. depressed with the sight of Africa again.'[1] His impression, not an uncommon one to travellers returning to Africa, soon evaporated and he remained on the continent for the following 17 years.

The next eighteen months in Bulawayo and in South Africa were a hiatus in his life. He lived the life of a 'gentleman of leisure', following his own interests. In Bulawayo, he stayed with friends and occupied his time perfecting his swimming technique in the local bathing pools, making bows and arrows, riding, birdwatching, and trying out various makes of motorcar. Just before Christmas, he wrote in his diary: 'Bought a new 6-cylinder 30 h.p. Chevrolet "Sports Roadster" car – a lovely 'bus with great power, very suited to this country. A Canadian product. Yellow and black, very striking and smart.' He was delighted with his purchase

– rather like a small boy with a longed-for Christmas present – and very soon after, he set off in his new car for South Africa.

Wood left Bulawayo in February 1933 to drive to Durban. It was the height of the wet season, and the roads were washed out and in a bad condition. At the border with South Africa, there were strongly enforced veterinary precautions owing to an outbreak of foot-and-mouth disease in Bechuanaland. But once in South Africa, Wood enjoyed driving on the good roads, although in one place there was an unbridged river drift where his car became stuck up to its axles in sand and muddy water. The Sports Roadster had an open canopy so, when it was fine, Wood could feel the sun and wind on his face, smell the aromas of flowers and trees and damp soil, and watch the birds. The verges of roads near the Limpopo were full of colourful flowers: a pale yellow ground-creeping flower with a spicy aroma, tall yellow evening prim-roses, bright mauve hibiscus, and purple foxgloves. After the Limpopo, he crossed the high veldt where cattle wandered all over the roads. It was the sort of country that he did not like: 'A howling gale in my face all today, and open prairie treeless country; cold, but sky cloud-less'. The only sadness was the dead birds: 'A tragedy of the high speed of cars on these roads is the number of swallows killed. The birds sit on the road and either don't move at all or only too late . . .' Wood was delighted to finally reach Durban on the coast: 'Timber and warmth again at last!'

In Durban, Wood settled down to the sort of life that he liked best. He had many friends there, and often accompanied them on expeditions. The days passed happily swimming and surfing in the sea, collecting seashells and butterflies, and practising archery. Wood loved the sea and the beach, and was always happy when close to the ocean and surrounded by verdant tropical vegetation. He was fascinated by the many shapes and patterns of seashells and took every opportunity to collect them. This fascination lasted for the rest of his life, and over the next thirty or so years, he amassed a huge collection of seashells from the coast of east and southern Africa and the Seychelles. He purchased a cabinet for his shell collection, and identified and labelled every shell with its full scientific name; and he made regular visits to the Durban Museum to obtain help in identifying his shells. He found the beaches and rocky coves around Umkomaas particularly to his liking and thought it was a place where he would like to settle. From the notes in his diary, it is clear that he spent many delightful days at the beach:

> Shell hunting again, swimming, etc. A lovely pool here cemented in rocks and flooded by sea at high tide. Warms up delightfully in the sun and one can bathe in modern costume and about stripped to waist! . . . Butterfly hunting, shell-hunting, bathing and archery daily now. There is a large patch of virgin jungle on north side of Umkomaas between railway line and sea almost extending to Ilfracombe which is really good for butterflies. Also contains monkeys and Blue Duiker.[2]

Wood also loved beachcombing and hunting for octopus. In August, he recorded the following: 'Hunting small octopods on rocks at Widenham, pushing them out of crevices as we did in the Seychelles. Masses of them are among the rocks now, many very small. Had splendid octopus soup for dinner.' And on another day: 'Hunting octopods at low tide in rock crevices; got 7. They appear to feed largely on crabs here. Excellent soup in evening.'

But sometimes there were misfortunes: 'Literally millions of "bluebottles" [stinging jellyfish] drifting in with strong easterly wind made all bathing impossible. Went to specialist about my ear trouble caused through bathing.' And two days later: 'After excellent surfing all afternoon at Durban beach through formations of a sandbank, was suddenly smitten down with ghastly sickness of severity have never had before; had awful time getting back to house in the car and very ill indeed all night. Blotto for three days . . . very weak and washed out; also my left ear very deaf and troublesome, and much catarrh about head. Fear I shall have to go very easy on this sea surfing and swimming for a bit.'[3]

After spending a short while in a boarding house, Wood moved to the Golf Course Club at Umkomaas 'on satisfactory terms' as a permanent resident. Later he was offered an appointment as secretary and overseer of the club in return for free board and lodging – a situation that suited him admirably and still allowed time for collecting and writing. Wood was still very interested and passionate about the usefulness of birds to humans, and so he accepted a suggestion from the *Natal Mercury* to write an article on the subject. This was printed on

17 June 1933, and included many personal anecdotes. After telling the story of the value of francolins in a field of maize (see Chapter 5), he continued:

> For years I was a planter in Nyasaland and made money out of it. For my own health and consumption, I grew countless fruits in that prolific soil. That fascinating friend of mine, the Bulbul (called toppie in South Africa) really did start annoying me by attacking my ripening peaches and guavas and mangoes. I did not rush to a gun. I did a bit of thinking. It was obvious that the birds had a mistaken idea that I was growing these fruits for them, so I produced a kind they liked better in sufficient quantity they would probably leave mine alone. As a result, I planted alternate rows of quick growing and free fruiting mulberry trees in my orchards. These served a double purpose, as they gave me windbreaks as well. My bulbuls were quick to recognise the fact that the countless thousands of mulberries must be their special perquisite – and for this small payment (I had many a mulberry pie too) they stayed ever with me, cleaning my trees of the harmful insects of all sorts which would have done ten times the amount of damage to my fruits than the birds could do. For years, I had so much fruit annually that I could supply all my friends and my labourers, and more fell and rotted on the ground. Of course, it was not only the 'toppies' that helped. There were many other kinds of birds too, but I mention them in particular as one so often hears maledictions poured on their cheerful black-topped heads! Long may they be with me in all future gardens I may have, with their ever cheerful songs and spirits when I am feeling glum – and they are welcome to their little payment of any kind of my fruit for all their help and encouragement to me.[4]

After citing several other examples of birds helping humans – Emerald Cuckoos destroying cotton stainer bugs in central Africa, chickadees controlling insects in spruce forests in Canada, owls feeding on rodents in croplands, pimento spice being propagated by birds in Jamaica, and many others – he continued:

> What I want above all these days . . . is sane judgement, the ability to observe correctly and weigh up the balance between good or harm. . .

On every wall in every school, and in every child's (and most grown-ups') bedroom, should be the slogan 'Protect our friends the birds'. In every school by every known means should a love of birds be engendered. Then perhaps we may see less of these senseless 'pellet-guns' and slaughterings, and more prosperous farms and orchards and agriculture throughout the land.

Two days later, the editorial column took up the theme and suggested that all gardens should have a birdbath:

So many people who have gardens do not realise how privileged they are. A birdbath (and it need not necessarily be more costly than any shallow basin or tin) repays its cost a hundredfold in the interest of watching all sorts of birds come to drink and stand in it awhile flourishing their feathers and getting as much enjoyment out of it as a man splashing beneath the shower on a warm day. . . . Every garden, however small, should be a place where there are birds as well as trees and flowers and where the birds know that they have friends. They very quickly become tame when they know there is nothing to fear and in their study one learns to know them as the most fascinating and beautiful of God's creatures.[5]

Wood's article was very similar in many ways to those he wrote for the *Nyasaland Times* – his writings show his frustration with people who could not see what was obvious to him, and whenever possible he made a heartfelt plea for the conservation of birds. As Wood commented himself: 'I hope it does some good.'

Although he derived a lot of pleasure from his Sports Roadster, it was not quite the car for bush-bashing. In fact, he found it very hard to know exactly what type of vehicle he did want, and he changed cars five times in two years. His second vehicle was a new Ford V8 Club Sedan which had the advantage that the seats could be adjusted to form a comfortable camp bed. But this vehicle was kept for only a few months, and then he exchanged it for a four-cylinder 24 horsepower Ford delivery van. He had the chassis reinforced and additional springs added so it could take up to a ton of luggage and be more appropriate for carrying all his collecting equipment. Then there was a V8 coupé suitable for the roads around Michaelhouse, and, finally, a V8 light delivery van, with a removable cover at the rear, which he drove back to Nyasaland in 1935.

Wood spent all of 1933 in and around Durban.[6] As well as going on swimming, surfing and collecting expeditions with friends and acquaintances, he made numerous lists in his diary of the species he had found. He kept caterpillars, and made notes on their development and on the plants that formed their diet. He went to plays and the cinema, and visited ships when they came to dock. On one occasion, when a French sloop visited Durban, he went on board and, with boyish naughtiness, quickly climbed up to the crow's nest to see the view; but to his disappointment he was evicted after being warned, 'C'est defendu là, Monsieur'! Towards the end of the year, he became interested in archaeological artefacts and stone tools, and started to make a collection of these as well. His many activities, and frequent changes in lodgings during the year, suggest that he was rather unsettled without any real challenges and with no direction. However, in June 1933, he met Kenneth Pennington who, like Wood, was an ardent lepidopterist. It was a meeting that changed Wood's life for the next few years.

The two men had very different characters, but both shared a passion for butterflies. Soon they were going on expeditions together. Pennington was born in 1897 (eight years after Wood), and his interest in butterflies was encouraged by his father, Archdeacon Pennington, who was also a keen naturalist and butterfly collector. Pennington was sent to an excellent school, Michaelhouse, located at Balgowan on the hills above Pietermaritzburg, where he received an education similar to Wood's. He was a brilliant young man: he read mathematics at the University of Natal at a very young age, and was awarded a Rhodes scholarship. But the First World War interrupted his studies; he became a pilot, had many daring exploits in England and the Middle East, and was awarded the AFC. After the war, he took up his Rhodes scholarship at Oxford, read law, and then returned to Pietermaritzburg to practise as a solicitor. But after two years, he heard there was a vacancy as a teacher at Michaelhouse and realised that his vocation was to return to his old school. He remained at Michaelhouse for 35 years, teaching mathematics and in later years becoming Senior Master and Acting Rector (Headmaster). During these years, Pennington was preparing a book about Southern African butterflies. This work, *Pennington's Butterflies of Southern Africa*, now revised and updated, is still the standard reference book on the subject.[7] Pennington realised Wood's worth and appreciated his vast knowledge of natural history and accomplishments as an archer, swimmer and surfer. So he arranged for Wood to come to Michaelhouse. It seems to have been a rather informal appointment, and there is no record at the school that Wood was ever a member of the teaching staff. But Wood himself was clearly delighted with the arrangement, and wrote in his diary:

> Decided to stay on here and help P. with Nat. Hist. Soc. work and also teach French, anyway for some time. Rector offered me quarters . . . in Common Room. Greatly bucked to have some aim and object in life once more: tones one up morally and physically and mentally. School life must be a high ideal and I am now of an age to appreciate that. So will see how it works.[8]

He moved into Tower Room by the belfry, describing it in his diary as a 'quaint spot under eaves of roof, but comfortable'.

Michaelhouse was (and still is) a private school where boys are taught to the highest standards, and are expected to do their best in all endeavours and be worthy members of society when they leave school. The school is set in a lovely parkland, with English-style stone buildings and courtyards, many spacious playing fields, and extensive views. Wood's duties must have been rather light, because there was time to go 'bug-hunting' each day when the weather was good, and to swim in the pool and exercise in the gym. He loved the school life, participated in lots of sporting activities, helped with lifesaving classes, took boys on bug-hunting and natural history expeditions at weekends, and had a great deal of fun. At the school, he was given the nickname of 'Archer Wood', no doubt because of his prowess at archery. In May 1934, the school magazine recorded: 'We welcome to the school Mr R.C. Wood, who is spending some months with us. He has been lending a hand in many useful directions, foremost among which is the assistance he has given to the already strong Natural History work of the School. "Archer" Wood is a very keen Naturalist . . .'.[9]

Under Wood's guidance, the school's Natural History Society had a large membership, and began to make collections of local birds, small mammals, reptiles and insects, all properly labelled and catalogued. Making good reference collections of local animals was one of Wood's special interests, and he imparted his enthusiasm for this activity to the boys. The museum gradually increased in size, and records of the local birds were made regularly. Wood gave talks to the society, and took groups of boys on birdwatching and insect-collecting trips close to the school. The society also became involved in archaeology, no doubt because Wood had participated in excavations at Mhloti near Durban, where artefacts of 'strandlopers' had been found. These strandlopers were early pre-Bantu people who lived on the beaches, probably 500–1,000 years previously, or earlier. Wood was fascinated by the pieces of pottery and stone tools that had been found, and his infectious enthusiasm for archaeological digs caught the imagination of some of the boys in the society. So when artefacts were found on Webster's farm, a few miles from the school, Wood sought permission to excavate there. At times he took up to 20 boys to dig and sift soil. The site proved to be a 'factory site' of Smithfield age (Late Stone Age); there were many broken artefacts, but also many in perfect condition. The then Director of the Government Bureau of Archaeology, Professor van Riet Lowe (a well-known expert at the time) visited the site and showed a lot of interest in the excavations. The society presented some 200 artefacts to the National Collection.[10] In all these activities, the boys learnt at first hand how scientific research should be conducted, how records and specimens should be kept, and how to identify the local fauna; and they had a lot of fun and enjoyment at the same time.

In May, the weather is cold in the Natal highlands. Wood was complaining that cold weather did not suit him and he was not feeling well. So when the winter holidays began, he set off in his van for the warmer climate of Nyasaland. Travelling by car in the 1930s was full of adventure, especially when the roads were bad. There was not a great deal of traffic in those days, but corrugations, potholes and bad edges often made driving hazardous. As he travelled, Wood, kept a note of every bird he saw along the way, and one suspects that his eye was only half on the road for much of the time. The worst part of the journey was encountered in Portuguese East Africa, where the road went up and down ravines and across streambeds. In one place, his car slid off hurdles placed into the water on a stream bed, and it took a couple of hours, with the help of local villagers, to get it on to dry land again. After a good night's sleep in a hotel at Tete on the Zambesi River, he set off again in the rain the next morning. The ferry over the Zambesi was pulled by a motorboat (cost 10/-). At the next river, his car was placed on a small punt and taken across to the other side. The road then passed through flooded *dambos*. All along the route, lorries were bogged down in the mud and water. Wood's car was stuck like this only once, and it was only because of good luck and the staying power of the car that he managed to finally reach Nyasaland. Wood recorded the distance and cost from Michaelhouse to Blantyre as follows: 'Total run: 1473 miles. 77 gallons of petrol = 19.13 miles/gallon. Cost £11. 2. 7d for all car expenses, plus £4 for pubs = £15. 2. 7d total.'[11] He spent most of his holiday seeing friends and bug-hunting in Zomba, Mlanje and Blantyre, but suffered from stomach aches – 'tum-tums' as he called them. During his time in South Africa, he had kept contact with many Nyasalanders and with those in government, and while in Zomba he visited the Chief Secretary and talked about the creation of the Monkey Bay Park (one of his projects when he was Game Warden that had not yet come to fruition). The return journey, in good weather, was uneventful apart from some eight punctures and burst tyres.

Wood remained at Michaelhouse for only two years, from 1934 to 1936, but during this short time he made a considerable and lasting impression on the school. Two of the boys at the school then were Vaughan Winter and Donald Currie. In 1995, sixty years on, as old boys, both retained vivid recollections of 'Archer Wood'. Vaughan Winter wrote:

> I think it was Ken Pennington who prevailed on Rector Currie to afford him digs at the School . . . He was given a small room on the first floor of the 'Clock Tower' to live, he took his meals in the Common Room, and used the showers in the 'old' West as his bathroom. No end of a moan if we left him no hotwater. I have an idea that he was educated at Harrow, as oft times you would hear Archer singing that Harrow was 'Forty Years On' in his shower.
>
> Archer was a great gentleman. A very sincere man with high princi-ples and ideals. A tall upstanding fellow and always neatly dressed,

whether roaming the hills or around the school. I may be wrong, but I don't think he found much in common with others on the staff. Although well read, and loved music, apart from K M P – and they must have found lots to talk about – I can't imagine Archer really hitting it off with the likes of —— and —— and the others of the time.

Archer was not formally introduced to us. He was an amazing man, and quietly merged himself into the School environment. The ice really broke, when he appeared on Aitkins one afternoon, carrying a boomerang in one hand and a lariat and a bow and arrows in the other. After watching a skilled display, and comments as he went through the various exercises in the use of the bow, boomerang and swirling rope, half the school was on Aitkins for the next performance.

Those of us who were interested in Nature and Wild Life soon found out that in Archer we'd found a man who was not only a tremendous source of information, but a veritable encyclopedia when it came to nature generally – whether animal, birds, butterflies and even archae-ology. He was terribly observant. Roaming the hills on free-bounds with him was an education. Sitting in his room after games, and paging through his books, or looking at photographs, or just listening to him talk, was always an experience, and a very worthwhile one too.[12]

After writing about the archaeological digs at Westfield, Vaughan Winter continued:

One of his achievements of course was to establish a School Museum. A small wood and iron building, at the top left hand corner of what is now Pennington Quad. It did not house much in my day. A few mounted local birds – Archer was no mean taxidermist, of course butterflies, and many of the stone implements we found on Westfield . . .

Archer looked after his health very carefully, and he was a stickler for exercise and keeping fit. On sunny days he would swim a few lengths at the swimming baths, then after more exercise, bask in the sun. Failing that, he would do a 'work out' in the gym with one of Bultitude's PT classes. I think he must have had malaria at some time. Sister King used to make up a concoction, which was basically quinine, and a cure for all ills whether sore throat, cold, 'flu, or any other ailment. There was always a huge jar of the foul tasting, cloudy liquid in each House wash-room.

Archer regularly took a draft of this, and said he preferred it to a glass of ale.

It was a sad day when he left. Meticulous in every way, we helped him carry his 'impedimenta' to his truck, but he would not let us pack it. Anyway, we gave him a great cheer as he left, and headed North to what was then Nyasaland.

Don Currie also retained very fond memories of Wood. He remembered that when he was a new boy at Michaelhouse, he caught a Death's Head Hawk Moth. At the suggestion of one of the senior boys, he took it to Archer Wood:

> Archer was excited by my capture and his enthusiasm fired my imagination. He took me under his wing and we spent many happy hours and days together catching butterflies and generally enjoying the wonders of nature. We often dug for archaeological treasures on the banks of nearby streams. All our efforts were made in order to build up an exhibit for our school museum . . . There is no doubt that 'Archer' Wood inspired my lifetime pleasure in the study of nature, and I am glad to say that I have been able to pass this on to my own sons and grandchildren.[13]

Wood always had a good rapport with young people, and it is surprising that he did not stay longer at Michaelhouse. He enjoyed his work at the school, and kept very busy, but he did not enjoy the cold weather in winter, when he had numerous colds and coughs. During the winter of 1935, he contracted pneumonia and was confined to the school sanatorium, and then took 2–3 months to recover completely. At some point during this period, Rear Admiral Hubert Lynes must have contacted Wood and made arrangements for Wood to accompany him on his forthcoming expedition to southern Africa. So in May 1936, Wood left the school for a last time and drove back to Nyasaland. His car must have been well loaded, for he had all his South African butterflies in their cabinets with him.

The return trip to Nyasaland was punctuated in Rhodesia and Portuguese East Africa with many stops to collect butterflies. Once back in Nyasaland, he had a busy time: he collected butterflies with his old friend W. A. Lamborn, watched birds, stayed with the Governor at Zomba and finalised plans for the Monkey Bay Park, and bought eight acres of land at Monkey Bay, where he intended to live. For a few weeks, he cleared bush around a possible home site, and planted trees. He decided not to build too close to the water's edge because of the probable rise in water level expected over the coming years. In this time, he also built a 'shack' – Wood always called his homes 'shacks' – where he could leave his 'impedimenta'. The District Commissioner, Mr W. H. J. Rangeley, sent him a pelican that had to be reintroduced into the wild, and for some time the pelican roosted on the roof of the shack. Wood also organised a new V8 truck for his forthcoming expedition with Hubert Lynes, and in early October set off once again for South Africa in preparation for meeting Lynes in Cape Town at the beginning of November 1936.[14]

Hubert Lynes was born in 1874.[15] He had an illustrious career in the Royal Navy, served in the First World War and was, towards the end of the war, captain of the battleship *Warsprite* which was present at the surrender of German High Sea Fleet on 21 November 1918. In 1919, when he retired as a Rear Admiral after 32 years service in the Royal Navy, he decided to devote the rest of his life to the study of birds. He had always been an ardent and knowledgeable ornithologist, and whenever he had leave from the Navy he went on birding expeditions. In his pursuit of birds, he was a prodigious traveller. Before retirement, he had studied birds in several Mediterranean countries, as well as China. After retirement, he visited Morocco on five occasions. He also went on two occasions to the Sudan, between 1920 and 1922, with Willoughby Lowe (mainly Darfur Province and Jebel Marra) where they collected over 3,000 birds, over 2,000 plants, 800 mammals and many insects. These collections resulted in a series of splendid papers which are still of great value for anyone interested in the biology of the Sudan. In 1930 Lynes started his travels and collecting in southern Africa, and on three separate expeditions he visited South Africa, Southern and Northern Rhodesia, Tanganyika and the southern Congo Basin. Although interested in all birds, he was especially interested in a group of small birds, the Fan-tailed Warblers, belonging to the genus *Cisticola*. These are small, brownish, rather nondescript birds which occur over a large area of Africa and which, along with many others, are sometimes referred to,

collectively, as 'LBBs' – Little Brown Birds – because they are so difficult to identify. In 1925, Lynes began an intensive study of this genus.
Although they are all LBBs, there are many subtle differences in colour
and pattern between species and geographical variants. When Lynes
began his study of museum specimens, there were about 173 specific
and 54 subspecific names attached to the genus. Trying to make sense
of this amazing complexity, he eventually classified the genus into 40
species with 153 geographical races. As an ornithologist, he was certainly
one of the best of his generation; he wrote many papers which were
published in *Ibis*, the journal of the British Ornithological Union, and
his review of the genus *Cisticola*, published in 1930, was still, at the turn
of the twenty-first century, the standard work on the subject. Lynes was
a very meticulous person who kept very precise and careful records,
and he made his plans and preparations with great attention to detail.
He was also very generous in giving money to expeditions and other
worthy causes. He was a great climber of trees and cliffs (a good skill
for an ornithologist), and he loved the challenge of finding birds' nests
and following birds through the bush.[16]

Lynes and Wood were an odd combination. In some ways they were
very similar: they both loved ornithology and the challenge of finding
out new information about birds, and both of them were well travelled,
tough, capable of spending long hours in the bush in pursuit of specimens, and meticulous in recording data. But in other respects they were
very different. Lynes had had an illustrious career, an unswerving devo-
tion to duty, and liked to be in command. He had great staying power
and would not suddenly give up and turn to something else. Birds were
his abiding passion, and he did not seem particularly interested in other
groups of animals. Wood, in contrast, had had one form of employment after another, and was quite happy to drift from place to place,
and from one activity to another. Staying in one place and doing the
same thing for long periods was not in his nature unless it was something that he really wanted to do, such as collecting butterflies and shells.
But he did have strong views on subjects which were important to him,
such as the conservation of soil, habitats and animals, and the importance of establishing good museums. By the time they set off together,
Lynes was 62 and Wood was 47. Lynes needed a younger man to help
with the driving, with the collection and preparation of birds, and with

the many small chores that needed to be done. Lynes' first impression of Wood when he arrived in Cape Town was very favourable. On 2 November, he recorded: 'Arrived Cape Town and alongside 7.30. Rodney W. came aboard about 9.00 having arrived from Nyasaland by van a week ago. All tip top. The Right man for me – charming fellow. We shall be good companions.'[17]

Lynes and Wood spent seven months together. Their itinerary was dictated by the places where they knew they would find particular species and geographical races of *Cisticola*. Lynes had done his homework very carefully, and knew exactly where he needed to go to collect specimens. During November and December 1936, they travelled slowly through Swellendam, George, Knysna, Port Elizabeth, Grahamstown and Bethlehem to Pretoria. In January 1937, they visited Swaziland, and in February they progressed down through Ladysmith to Durban. During two visits to Pretoria, they examined specimens in the Transvaal Museum with Austin Roberts (who wrote the standard book on South African birds) and van Someren (a foremost authority on African butterflies). On the second visit, in February 1937, Lynes had to spend several days in a nursing home recovering from an unknown malady and from exhaustion.[18]

Lynes and Wood both kept diaries during their travels and, as is to be expected, each recorded slightly different things about their travels. For each day, Lynes recorded (in blue or black ink) how he had slept, the maximum and minimum temperature, when he got up and went out, and his impression of the weather. For example:

> Mon 14 Dec. PE to Grahamstown. A decent night, out at 4. Lovely calm blue sky, streaky dawn. Temp 4.00 min 56°. Barom (went down 3/10 yesterday) 30.2.
> 30 April. Kisumu. Sound asleep 9.30 to 2.30. Out at 3. Temps 3.30 67° 65°. Calm. Fine.

This was followed by notes on birds, what they had collected, and where they had travelled. Lynes also added, in red ink, the names of all the people to whom he had written letters that day and a few words on the subject of each letter.[19] Every day, after working in the bush, he and Wood had to prepare and label the museum specimens of all the birds they had collected. During their travels, they stayed at hotels, but since they were moving every few days, they were constantly packing and unpacking. They found each day to be long and tiring.

Travelling was not easy, mainly because of the state of some of the roads and the weather. Nevertheless, they continued whenever possible to collect specimens of *Cisticola* and butterflies. Wood's diary in early January 1937 gives some idea of the problems they encountered while returning from the coast near Kosi Bay:

19 Jan. A ghastly day, back thro' sand and great heat – cars boiling every few miles, all way, labouring in sand etc. with thermometer well over 100°. At Pongola R. troubles ended and we had crossed by 4 pm. In tropical forest along bank (east) of river, *Pieris spilleri* very numerous indeed. Up Lebombo Mtns to Inhwavuma and onto Gollel Hotel (good) by 6.45 pm. Absolutely played out.

20 Jan. Standing easy at Gollel in great heat again. Heavy storm in evening.

21 Jan. Tried to get along to Stegi, but found Inhwavuma R (20 m. N) in full flood and had to return to Gollel. Ditto on 22nd.

23 Jan. Got across Inhwavuma and Hlatuzi rivers at last and so to Stegi.

24 Jan. Stegi. Weather and roads too bad to get down to Ndumu.

25 Jan. Along edge of Lebombo Range. [Here RCW lists all the butter-flies he caught]

27 Jan. To Machadodorp, via Carolina. Roads in places dreadful . . . bogged on flooded spruit near Machadodorp where luckily locals and other people hauled us out much mud-covered.

28 Jan. On highest bleak hilltops 8 mi along road to Lydenburg, where found only *Cisticola ayresii* in some numbers.

29 Jan. After waiting until noon, crossed flooded river in front of Hydro Hotel, and then had awful run through rain and mud, being finally held up 7 mi from Middelburg by deep flooded spruit, and so had to return some 38 mi to Belfast where stayed at excellent Transvaal Hotel.

30 Jan. At Belfast all day, held up by weather and roads.[20]

During their travels, Lynes and Wood had to carry their precious spec-imens with them, carefully packed to avoid damage from bumps, insect pests and excessive humidity. Every specimen was collected for a specific reason – to show geographic variation of a particular population, or to prove that the species (or subspecies) occurs at a stated location, or to show dimorphism between males and females or between adults and juveniles. To scientific collectors such as Lynes and Wood, care of every specimen is essential to justify its collection in the first place. When slushing around in mud and water, with rain or hailstones beating down upon them, they must have been very concerned for the safety of their collections.

In early March 1937, Lynes and Wood, together with their van, boarded the SS *Tairea*, an 8,000-ton twin-screw steamer of the British India Steamship Company, for the sea journey to Kenya. The steamer stopped at Lorenço Marques (where Wood had time to go ashore and catch butterflies on one of the local beaches), and eventually reached Mombasa nine days after leaving Durban.[21] The plan for the next three months was to travel through Voi, Nairobi, Kinankop and Eldoret to Kisumu on the shores of Lake Victoria, and then to take the boat down the lake to Mwanza and drive though southern Tanzania to Nyasaland, from where Lynes would travel by train back to Cape Town. On arrival in Mombasa, Lynes hired Hamis bin Ali as his *safragi* (servant on duty) until they reached Nyasaland.[22] Hamis came with impeccable references, including one from the Prince of Wales. It took two days to sort out all the red tape and legal formalities, and eventually Hamis was signed on at a salary of £3 per month plus 1/- per day for *posho* and 30/- for his outfit (blankets, shorts, shirts, umbrella or whatever else he wanted to buy). Lynes had to undertake to return Hamis to Nairobi at the end of the journey.

This was the first time that Wood had visited East Africa (other than the port of Mombasa), and he delighted in the scenery and the animals. While staying in the Namanga Camp Hotel, 83 miles north of Arusha (close to what is now Amboseli National Park), he went off in search of butterflies, as recounted in his diary entries: 'On 23rd, out in forest, came across some elephants resting in streambed, much rhino spoor, two impala and one or two dik-diks. In afternooon while trying to approach a giraffe to photograph him, a lone elephant slunk off close by. This place full of such animals. A tame ostrich has eaten a packet of nails two days ago!' And: '24th. Out again in bush to east of hill; rhino spoor, bull giraffe, Tommies, granti, 8 giraffe, 4 crowned cranes, large herd impala. Kilimanjaro out of cloud, magnificient!'[23]

The itinerary, as in South Africa, was planned so that visits could be made to all the important *Cisticola* localities. However, both men were by now suffering from travel fatigue. Soon after getting off the boat, Lynes recorded: 'R was upset in the tummy all day and has become "very irregular"', and put it down to the sea voyage. 'I think he goes too heavily for purgatives and thinks too much of "my cabinet". R absent from dinner and I dined by invitation with the nice Andersons who are returning to South Africa.'[24] There were numerous problems as they travelled through Kenya, as well as punctures, rainstorms and muddy roads that caused many delays. But the real reason for friction, according to Lynes, was because Wood was tired of collecting *Cisticola* and was interested only in obtaining certain butterflies. By the time they reached Iringa, in central Tanzania, Lynes was frustrated but philosophical. He knew that illness, stress and bad weather can fray tempers very easily. On 1 May, he wrote:

A good night, 9.30 to 3.00. Out at 3.30 feeling much recovered from — — . . . R. has really lost all interest in Cisticolas and it's become rather hard effort on his part to help me as he did in S. Afr. And wishes me to understand by the scarcely indirect question 'What sort of a bird is emini' after we've been hunting for teitensis so long – after I've done my best for him with Cist . . . So we must modify programme – certainly cut out Mweru and Bangweola and hasten back to Monkey Bay. The great thing for me to remember is to acknowledge with gratitude his help in S. Afr with Cisticola there . . . and with the van. He is now with the road travel almost hysterically pessimistic like he was . . . at the prospect of the Nyansa trip when R. said there were storms in the rainy season. 'Oh this is impossible for me. I shall be seasick the whole time and probably be buried at Mwanza. We must try to get by road Kisumu to Mwanza by Uganda.' Well! Be calm, considerate and tolerant. Less than a month of May remains.[25]

A few days later, Lynes confided the following to himself in his diary, after the normal comment on sleep ('a grand night 9.30 to 3.30, out at 3.45') and the weather:

R has lost all interest and . . . toleration for Cisticola things. . . . So I thought it out and in future I'll keep off dangerous ground

147

(for he is always right and tells me I am always wrong – that fatal self-centre-ism) and go out without him.[26]]

Both Lynes and Wood were very well-educated men, and both liked good conversation and humour, but they had a habit of baiting each other. Reay Smithers, in a paper titled 'Recollections of Some Great Naturalists', recalled what it was like travelling with them:

> The diverse interests of Lynes and Rodney, in spite of their deep friend-ship, led to many good-natured but nevertheless noisy arguments in respect of the merits of collecting small brown birds as opposed to butterflies. Lynes was usually the first to start these arguments, and mischievously inclined, would say in his quiet way, 'Rodney, why do you spend your time collecting butterflies – what possible value is there, it's just like collecting postage stamps, it's obviously simply because they happen to be prettily coloured, in fact, they have little scientific value!' Rodney would reply that Lynes ought to be ashamed of himself, 'spending his life killing small birds from one end of Africa to another', and they would be off.[27]

When they reached Mwanza, they visited the shore of the lake close to the village of Busissi where Emin Pasha had obtained the type specimen of a *Cisticola* at the end of the nineteenth century, and which was later named after him as *Cisticola emini*. Emin Pasha was one of the most brilliant and enigmatic of the Europeans who lived and worked in Equatoria and around the source of the White Nile during the last years of the 1800s; like many of the other dedicated doctors, administrators and military men who lived in Africa during the early years of colonialism, he collected many specimens of animals, which eventually found their way back to Britain. For Lynes, it was essential to find some specimens of this species at the type locality; these would be as close to identical as possible to Emin Pasha's specimens and hence show the definitive characteristics of the species, without any geographical variation that might be present elsewhere. For several days, Lynes and Wood hunted – successfully – for *emini* in the low bush-covered rocky hills near the shores of the lake.[28]

Most of the roads in central Tanzania were impassable because of the rains, so they put the van on the railway to Tabora. Here, they had a pleasant break when they stayed with the 1st Batallion of the King's African Rifles and were able to celebrate the coronation of King George VI on 12 May. They continued by train to Dodoma, but then were able to take to the roads again. The weather was wet and cold, especially in the Southern Highlands. The road crossed many stream beds and drifts, all of them rather hazardous. In one place a small bridge broke while they were crossing it, but luckily there was just enough time to accelerate to the opposite side before it collapsed into the deep pool below. Ten days after leaving Dodoma, they arrived at Monkey Bay, probably thankful that the trip was over. Wood accompanied Lynes to Blantyre, where they stayed in Ryall's Hotel for a couple of days. Lynes then travelled by train to Johannesburg and so home to England, and Hamis, the *safragi*, returned to Nairobi. On the day when Lynes left, Wood wrote: 'Lynes left by train for the south, leaving the car with me as a present. An amazingly generous man. Thus ends our trip together – very enjoyable and with good humour and friendship throughout.'[29]

Wood spent most of the following year, from June 1937 to July 1938, in Nyasaland. To begin with, while he was building his house with locally made bricks, he lived in his 'shack'. By September the roof was on, and in October, just before the wet season began, he moved in. He went bug-hunting whenever possible, and planted many trees and bushes around the new house. In early February, he spent three weeks on Chingozi Estate, close to the southern escarpment of Mlanje Mountain, and not far from his old home at Magombwa. One day, as he was walking on the forested lower slopes, he collected several specimens of a small butterfly, not much more than an inch across the wings, yellow with a black border on the upper side and yellow with black lines and spots on the underside, which he recognised as being unusual. In his characteristically careful way, he recorded in a letter to N. D. Riley at the British Museum of Natural History that all the specimens were taken in the same place: 'a small vegetation-clad rocky islet in the main stream-bed of a forest-fringed large mountain river on the side of Mlanje Mountain at about 2500 ft. They sit on the thin stems and trailing loops of lianas climbing or festooned from tree and bush, or on the terminal twigs of dead branches of the bushes beneath the tall forest canopy. Flight is sluggish and weak, as is characteristic of the group.'

They proved to belong to a new species of lycaenid butterfly, and were named *Teriomima woodi.*[30]

In late February 1938, Wood was on the move again – maybe because it was necessary to earn some money to keep body and soul together. He travelled on the SS *Mpasa*, one of the African Lakes Corporation steamers, from Fort Johnston (where the Shire River leaves the lake) to Florence Bay (on the western side at the northern end of the lake). From here he travelled to the headquarters of the Scottish Mission at Livingstonia.[31] The mission was established in 1894 at Livingstonia by Dr Robert Laws after the original sites near the lake, at Cape Maclear and Bandawa, had proved to be unhealthy (see Chapter 2). Livingstonia is situated in the hills, high above the lake and immediately below the Nyika Plateau. It is a spectacularly beautiful place, but can be cold and wet at some seasons of the year – perhaps rather reminiscent of Scotland to Laws and his missionaries. Wood was employed by the mission to teach agriculture and soil conservation, subjects that he was well qualified to teach, and to start a museum.[32] It seems an odd appointment because Wood was not a religious person, and he had a rather low opinion of clerics and missionaries. He probably thought that teaching conservation and natural history to local Africans was so important that he was prepared to put aside any differences in religious outlook between him and those at the mission.

Soon after arriving at Livingstonia, he spent a week on a sailing trip with the Reverend W. C. Galbraith from the mission.[33] It was still the wet season – rainstorms and cloudy skies were common, and at times the water was rough. During the day, they rowed and sailed southwards, stopping every so often so that Wood could go ashore and collect butterflies. On one day, they watched a huge black cloud of Kunzu flies passing along the lake (a well-known sight there); but on this occasion, the flies were being pursued by 'scores of thousands' of swifts or swallows which were feeding on them. Wood commented that he had never seen such numbers before and that they 'looked exactly like a swarm of locusts'. At night they camped on the shore. One night was especially memorable:

> Heavy storm at night (about 4.00 hrs). Brought the tent down. Struggled in to some clothes and crawled out of wreckage. We felt way to local

council hut some 100 yards away by feeling edge of path with our feet and taking direction by lightening flashes. Took shelter there, rather wet, until dawn. Rain continued until 10.00 hrs when we got along to tent wreck. Everything, bedding, mattresses, etc. wet through. Spent day trying to dry out bedding but hardly any sun. Got two blankets, cork mattress and pillows ± dry, so very lucky.

After that soggy experience, they returned (quite sensibly) to Livingstonia, where Wood spent the next few months teaching agricultural supervisors.

After all their differences on the first expedition, it is perhaps rather surprising that Lynes kept in touch with Wood, and asked Wood to accompany him on a second expedition.[34] However, Lynes knew that Wood was an excellent person to have on an expedition – knowledgeable in so many ways – and that he probably could not find anyone more suitable. Lynes was determined to go to South West Africa to find *Cisticola subraficapilla windhoekensis*, a geographical variant only found around Windhoek, and to make further collections in South Africa and further north. So in late July 1938, Wood travelled by train and truck to Cape Town, where he bought a Chevrolet Sedan van, and a special trailer to hold all their luggage. Lynes arrived on the Union Castle liner SS *Bloemfontein* on 15 August. Within a few days, they set off towards Port Elizabeth and Grahamstown, and spent a fruitful two months collecting *Cisticola* specimens. On this trip, they took Ali Safi, of the Cape Town Museum, who was hired to prepare the specimens of the birds they collected; this meant that they could collect more specimens and make more observations than would have been possible otherwise. On their return, they decided that the van and trailer were inadequate for their needs, so Lynes purchased a Chevrolet ¾-ton pick-up truck. In early November 1938, they left Cape Town for South West Africa, Ali driving the pick-up and Lynes and Wood driving the van. The roads were appalling. On 3 November, Wood wrote that the 'road [was] awful with corrugations all way; body bolts of pick-up all shaken loose and much trouble. Rain all day.' On 6 November, they arrived in Windhoek 'after days of the most incessantly awful road I have ever been over in Africa; endless shaking and vibration and endless corrugations making wear on cars appalling'. However, their spirits rose on the following day, because on a nearby farm they found many of the *Cisticola* that they had come for, and they were able to collect several specimens. They could not face the idea of driving back along the terrible road, so they, together with their cars, took the train back to Keemansshoop, where they found another local form of *subraficapilla*. Finally they crossed the Orange River on a punt, and reached Springbuck. By this time, Wood had seen enough of South West Africa: he had caught a chill because of the cool winds, the roads were bad, the country was 'frightful' and there were no butter-flies; his final valediction was: 'Anyone can have it!'

Lynes planned that the second part of the trip would be along the coast from Cape Town to Port Elizabeth and Durban, then to Balgowan and Pretoria, and finally through Southern Rhodesia to Nyasaland. They took about a month to reach Balgowan, where Wood stayed at Michaelhouse with Pennington, and Lynes stayed near by with Jack Vincent, the herpetologist. However, the rest of their well-laid plans did not materialise. Wood had numerous tummy upsets and was irritable, and Lynes suffered from kidney stones. Worst of all, Lynes developed a bad eye (because of a herpes virus) and decided in Pretoria that he would have to abandon the trip and fly back to Britain. He was keen for Wood to continue collecting, and offered to pay all expenses. Wood and Ali took the two vehicles to Southern Rhodesia, but could go no further than Salisbury because the wet season had started and all the roads to Nyasaland were impassable. Wood and Lynes remained in contact by telegram (or by wire, as it was called then), and decided that the expedition had to be cancelled. Ali returned to South Africa with the pick-up, and Wood took the car (donated to him by Lynes) by train to Beira and on to Blantyre. Lynes ended up having to remain in Johannesburg for six trying and painful weeks, and eventually reached England in February 1939. So ended the expedition that had started with such high hopes.

Lynes had planned to write an appendix to his 1930 *Cisticola* review, based on the material collected during his two expeditions with Wood. But by the time he was well enough to embark on this work, the Second World War had begun. He was determined to serve his country again, but his applications for war service in any capacity were turned down. Eventually he was given a short-term appointment in the Navy, a posi-tion to which he applied his usual vigour and professionalism, but this only lasted until the end of 1942. On his 'retirement', he was about to start on his *Cisticola* work when he died suddenly after a short illness,[36] and so the appendix to his review was never written. However, all the specimens collected by Lynes and Wood are still carefully preserved in the Bird Room of the British Museum of Natural History, and are available to any ornithologist who wishes to study them. As with Wood's collections, they are a permanent record of where particular species and geographical races were present at various times in the 1930s, and are an essential source of information for any reviews, now or in the future, on this group of birds. It is not easy when examining specimens of birds, mammals or butterflies (or indeed any animal or plant) in a museum to realise the immense effort and sacrifice that had been necessary to collect each specimen. Lynes and Wood worked very hard and with great dedication to obtain their specimens; they had constant problems with roads and vehicles, they were tired and ill on many occasions, they had arguments and differences of opinion, and yet they continued on in their pursuit of knowledge. To me, and to many other biologists, it is a magical experience to examine a specimen, still in fine condition, and read the original label in the collector's handwriting – and to try to imagine what it must have been like fifty or a hundred years ago when the specimen was collected, and when life and travel in Africa was so different to what it is today.

Wood returned to Livingstonia, where he continued to train agricultural supervisors and collect specimens for the museum. He did not like the cold, cloudy and wet climate of Livingstonia or the lack of butterflies except when it was fine and sunny, but he still retained his intense curiosity about the natural world around him. At one time, he had a tame crowned crane, which had been brought into the mission as a youngster. The most distinctive character of these cranes is a fluffy tuft of golden-coloured feathers on the top of the head. Both sexes have the tuft; its function is uncertain, although it is often assumed that it is a secondary sexual characteristic and used for display. Wood kept the bird, which he guessed was a female, although there is no difference in plumage between the sexes, in a large grassy enclosure at the side of his house where the local grasses were not cut. One day he went to look for her and was unable to see her. Eventually he found her, right under his feet. She had settled on the ground with her feet well tucked under her body and with her neck and head buried under the blackish-brown feathers of her back. The feathers of the crown were projecting from the back feathers, and rustling in the breeze. He wrote:

> Never have I seen a better piece of mimicry. Her whole outline, low and depressed on the ground, was entirely broken by the gently stirring 'grass-tuft' and its shadow in the sunlight. She was safe from any great eagle or other enemy from the air while in one of her siestas of deep sleep in the heat of the afternoon. From the side, the grass around her as she lay flat covered her very effectively, and the 'crown' became even more obviously a tuft of grass-like vegetation.[37]

Wood concluded that, to him, it was more likely that the crest was for camouflage than for display. At another time he was given a baby Crowned Hornbill whose feathers were just beginning to sprout. Wood described his baby as 'an incredible looking thing; shall try to rear it with much pleasure. It eats banana, grasshoppers (with great gusto) and cold pieces of chicken.' As anyone who has reared baby birds knows, it can take a considerable time to find suitable foods and to feed them, but the result is always worth the effort. Undoubtedly, Wood must have had a lot of enjoyment looking after his baby hornbill.

He also became interested in the Vulturine Eagle. This eagle was known to feed on oil palm nuts rather than on flesh, but it also occurs in areas where oil palms are not present. While living at Livingstonia, Wood discovered one in a raffia palm tree near the Stone House, formerly the home of Dr Robert Laws, the founder of the mission. Wood's careful observations are recorded in a paper he wrote in *Ostrich*, the journal of the South African Ornithological Society, in 1943:

> . . . I noted in my diary that a Vulturine Eagle had been living in the trees around the Stone House where I was then staying . . . and I had twice flushed it from a solitary Raphia Palm which grows at the edge of the croquet lawn, some ten yards or less from the house door. For some time, I had noticed that some animal was dropping palm fruits [from which] the pericarp had been gnawed away . . . I suspected some kind of squirrel although I had never seen any squirrel hereabouts. This particular plant

is . . . fully mature with gigantic bunches each containing thousands of tightly packed fruits.

About this time, other members of the staff had remarked to me about this bird which they had never noticed before, being flushed from the palm when they came along to play croquet. While the croquet was in progress (and what a noisy argumentative game it is!) the bird used to circle around or sit in some tall citriodora eucalyptus trees nearby, waiting to get back to the Raphia Palm.

I therefore started to watch more closely. On 13th May, I noticed that the Eagle had been around daily and . . . I felt convinced that it was feeding on the palm fruits. Knowing the bird to be in the tree, although invisible among the dense mass of foliage and fruit bunches . . . I watched fresh fruits falling to the ground and prompt examination of them while still wet and juicy showed distinct beakmarks all over the seed where the outer yellowish pulp, the pericarp, had been pulled off. Each time I came near for these examinations, the bird flew out of the palm. By 17th May, I was absolutely convinced that this bird, and no other animal was responsible. It never made any attempt to eat the young kernels of the nuts even when these were still soft and easily opened up. I may add that the nearest oil palms are in a small grove on the shore of Lake Nyasa a few miles away from Florence Bay. I was able to examine fallen nuts from them which had been eaten in the same way, i.e. the pericarp only.[38]

Another of Wood's interests during these years was finding archaeological artefacts, as described above. At this time, very little was known about the archaeology of early humans in southern Africa, so any stone tools, bored stones and pieces of pottery were of interest. Now that he was back in Nyasaland, he hunted on river banks and on eroded lake beaches near Manchewa and Mwenerondo (south of Karonga). Some of these old beaches were 100 feet above the present water level, at a time when the lake was much more extensive than now. He found numerous Stone Age implements here, especially rough flake tools and 'splendidly' bi-faced tools. Many of these sites became famous in later years as important sites of early human habitation in central Africa. Nearby also, at Chiweta, are important fossil beds containing the remains of dinosaurs where he found some bone fragments.[39]

Soon after Wood's return to Livingstonia, the political situation in Europe deteriorated rapidly, and within a few months Germany invaded France and Poland, and the Second World War began. Nyasaland was not affected to the same extent as during the First World War because the former German East Africa was now Tanganyika and part of the British Commonwealth. However, Nyasaland was a staging post for troops and aircraft travelling north from southern Africa to Ethiopia (for the campaign against the Italians) and to Egypt (for the desert campaign in North Africa). The Nyasaland battalion of the King's African Rifles was mobilised, and saw active service in Ethiopia and Burma. Wood was too old to enlist (he was now 50), but he spent a lot of time growing vegetables near the mission for the KAR troops stationed at Ekwendeni, and he purchased Nyasaland War Savings Certificates to help the Nysasaland war effort.[40] Nyasalanders were able to read about the war in North Africa and Europe in the *Nyasaland Times*; each issue was 8–12 pages long and cost 3d; the paper was printed twice weekly but on poorer quality paper than previously. Radios were freely available, and Wood bought himself a Phillips radio, powered by a 6-volt accumulator battery, so he could tune into local stations which relayed the BBC World News. Despite the problems associated with the war, he still collected butterflies and watched birds when the opportunity arose.

Nyasaland had changed a great deal over the last few years, especially with respect to transport and communications. Cars were now the main means of transport. In 1939, a Hillman Minx saloon car could be purchased for £220; such a car (as the advertisement said) had 20 new features and had remarkable economy – over 40 miles per gallon.[41] Cars were no longer advertised in the local papers as often as they had been in the 1920s, perhaps because they were by now so commonplace. The main roads were much improved, often tarred (sometimes with just two tarmac strips for the wheels), and travelling around the country was quicker and safer, at least in the dry season. The railway had not changed so much; in the early 1930s, the line was extended from Blantyre to Salima, a few miles from the lakeshore, so that travellers could now go from the north to the south of the country using only steamers and trains. In 1935, the huge bridge crossing the Zambesi at Mutarawa was opened, providing a direct train link from Nyasaland to the port of

Beira. The bridge was one of the most spectacular engineering enterprises in Africa. It was 4,000 metres long, with 34 spans across the river
itself and 47 altogether.[42] Passengers no longer had to take the overnight
ferry across the river, but could stay on the same train for the whole
journey. The Blantyre–Beira train left on Sundays and Wednesdays at
10.50 a.m., and arrived in Beira on the following morning at 8.25 a.m.
But the most significant change in transport was the introduction of
air travel to Africa, although in this respect Nyasaland was a backwater
until the end of the war. The advertisement for Imperial Airways in the
Nyasaland Times of 1939 shows that air travel was delightful in those
early days:

> Less than six days from Durban to England. It is the luxury which
> surprises so many travellers in the four-engined Imperial Flying Boats.
> They find themselves taking meals in the comfortable soundproof salons,
> enjoying the passing scenery from the promenade deck, smoking or having
> their afternoon nap, with a steward always at their service. Nights are
> spent quietly on land in hotels and resthouses. The fare includes every-
> thing including tips. Try this luxurious method of travel. A single expe-
> rience will make you an enthusiast.[43]

However, most people travelling between Nyasaland and Britain still
used the many ocean liners which departed from Beira, Durban or Cape
Town. The companies and the liners were almost household names in
those days. Best known was the Union Castle Line, whose boats were
called either after famous British castles, such as *Stirling Castle*, *Arundel
Castle*, and *Carnarvon Castle*, or after places, such as *Bloemfontein Castle*,
Capetown Castle and *Pretoria Castle*. There was also the Ellerman and
Bucknall Line with its City boats – *City of Paris*, *City of Exeter* and *City
of London*. Passengers destined for Europe travelled by either the Holland
Africa Lyn to the Netherlands, or by the Deutsch Oest-Afrika Line to
Hamburg. The cheapest single fare to London was £40 from Cape Town
and £50 from Beira.[44]

Radios were widely used throughout Nyasaland by the end of the
1930s. The programme for 'Transmission 4' was printed in each issue
of the *Nyasaland Times*. Transmission started at 7.25 a.m. with five minutes
of 'Sports News and Market Notes', then there was half an hour of

music, followed by 'Big Ben and the News' at 8.00 a.m.. Between 8.15 and 12.30 there was music or opera, plays and humorous programmes ('More Food for Thought', 'I want to be an Actor', 'Time to Laugh', 'Entertainment from Blackpool'), and commentaries by Sandy Macpherson. Transmission lasted for only about five hours each day.[45]

Postal services continued to improve every year, and post offices were established in ever more remote areas. For Europeans, the greatest advance was the beginning of airmail services to Britain and other parts of the world. The airmail service to Britain departed from Blantyre at 11.00 a.m. on Wednesdays and Saturdays, arriving at Southampton at 12.40 p.m. on Tuesdays and Fridays. Airmail services to East Africa, South Africa and the Belgian Congo left at the same time. There was even an airmail service twice weekly within Nyasaland between Blantyre and Lilongwe, with a delivery time of only five and a quarter hours. Prior to the 1930s, telegrams (or wires) were the quickest way of sending messages from place to place. But gradually telephones were introduced, at least in major towns. Telephone numbers were very simple: one advertiser in the *Nyasaland Times* gives the number as just Blantyre 34.[46]

Wood travelled extensively in his car, although he also had a Hercules bicycle for short trips. He changed his car several times after his return to Livingstonia in 1939 at the end of the second Lynes expedition. It is difficult to imagine why he made so many changes, other than for reasons of economy and because of what was required at the time. He sold the Chevrolet car which he and Lynes had used in South West Africa and South Africa, and bought a small DKW; but within a few months, he swapped this for a Chevrolet Coupé Imp, which he kept for two years. On his return to Monkey Bay, he sold the Imp and bought a 2-seater Morris 8.[47]

By 1942, Wood could not stand the cold, damp climate of Livingstonia any more – it was a 'depressing place' by his standards. So on 24 May 1942, he packed his belongings into his car and set off for Monkey Bay: 'Arrived on 28th, after staying with Lamborn at Ft. J. as usual. All in good order and trees around house grown mightily. What joy to be back again!'

Chapter 8

LIFE BY THE LAKE: 1942–1956

Monkey Bay and Cape Maclear are situated on a large peninsula which juts out into Lake Nyasa at the southern end of the lake. The lakeshore is a mosaic of rocky outcrops, sandy beaches and sheltered bays. Inland from the coast, the vegetation is dry savanna on rather sandy soils, with rocky kopjes (or inselbergs) scattered here and there. The most notice-able feature of the vegetation is the large number of baobab trees: huge primeval-looking trees with broad trunks and wide open branches. They look so different to other trees that, jokingly, it has been suggested that they have been planted upside down. Each baobab is strikingly different in shape and form from any other, and is easily recognisable. In the dry season, the branches are bare, producing a lovely tracery against the bright blue cloudless sky. Monkey Bay is to the east of the tip of the peninsula; it is a tranquil place, only about a mile across, and protected from the winds and weather of the lake by rocky headlands. It gained its name from all the baboons that used to live in the rocky savanna hills surrounding the bay. The little township of Monkey Bay is on the southern side of the bay; it looks northwards towards the headlands and to the lake beyond. A few miles to the west, on the other side of the peninsula, is a long, beautiful sandy beach. Here at the eastern end is a local village, Chembe, and at the opposite end is Otter Point, named after the lake otters that can often be seen catching fish in the crystal clear waters, as well as another headland called Cape Maclear. It is a place of great importance in the missionary history of Nyasaland, because it was here that Dr Robert Laws built the first Livingstonia Mission in 1875 (see chapter 2).

Wood spent 16 years of his life, on and off, in this quiet off-the-beaten-track part of Nyasaland where the tempo of life was governed by the moods of the lake and the alternating contrasts of the wet and dry seasons. At first, from 1942 to 1950, he lived at Monkey Bay, and then, from 1950 to 1958, at Cape Maclear. Apart from visits to the Seychelles (see Chapter 10), he left the peninsula only when necessary. On two occasions he lived near Mulanje, where he supervised tea plantations: from October 1942 to February 1943 he was at Chingozi Estate, and from November 1944 to June 1946 he worked at Mimosa Estate. He did not like living near Mlanje – there was incessant rain (which is why it is such a good place for tea), and he intensely disliked the cold, changeable weather. But these periods of employment were necessary to supplement his income; for the rest of the time, he remained on the Monkey Bay peninsula and was able to live the simple life, following his own interests, from which he got so much enjoyment.

On arrival at Monkey Bay in May 1942, at the beginning of the dry season, he started to build the second part of his cottage and the quarters for his staff which he had started before going to Livingstonia. He arranged for 40,000 bricks to be made,[1] a task that can only be done in the dry season. In Nyasaland, bricks were made (are still made) by stuffing clay into brick-sized wooden moulds. The moulds are then removed, and the bricks are stacked over piles of firewood and covered with soil. The firewood is lit and smoulders gently for one to two weeks, by which time the bricks are fired and ready for use. It took only a few months for Wood to complete his cottage. He did much of the work himself because he derived considerable enjoyment and satisfaction from this sort of Robinson Crusoe 'do-it-yourself' lifestyle. He managed to complete everything before going to Mlanje for a few months in October 1942. During his absence, and during many other absences over the coming years, his staff looked after the cottage and garden. He was always delighted to return to this remote quiet haven. He used to record his pleasure in his diary: 'Everything green and more beautiful than I can express. My trees and shrubs growing famously. What joy to be back!'[2]

Life in Monkey Bay and Cape Maclear was, for the most part, simple and enjoyable, consisting of visits from friends, collecting butterflies, watching birds, tending the garden and enjoying the changing scenes on the lake. But at times there were frustrating problems, such as the constant malfunction of engines and pumps, the bad state of the roads in the wet season, floods, the menace of leopards, baboons and snakes, and various illnesses. For much of the time, nothing much happened on the Monkey Bay peninsula; the only exceptions were the 'Vipya' disaster and when the Flying Boats landed at Cape Maclear. But Wood's life at Monkey Bay in these years provides an interesting snapshot of what it was like to live up-country in Nyasaland in the 1940s and early 1950s.

Wood's cottage was quite small – a living room which opened on to a *khonde* (veranda), a large bedroom which also served as a study, a spare bedroom, a bathroom and toilet, and a store and ironing room. From his *khonde*, Wood looked out over the bay and to the lake beyond. The kitchen was in a separate building some distance away. In those days, kitchens were never, or rarely, part of the house. The garden outside was full of trees and flowers that attracted many birds. There were bird-baths, much frequented by many species, and feeding tables which attracted them close to the house. There was a vegetable garden, and chickens that provided a supply of fresh eggs. Wood employed several local villagers to look after him. Wages were low and even people with limited resources like Wood could afford to employ a number of staff. Very often, the same people were employed for years and years, and they became almost part of the family. It was beholden on employers to provide their staff with more than just the monthly wages – uniforms, schoolbooks for their children, medicine, extra money to buy thatching grass for an employee's house in his village, and perhaps a bicycle so that the cook could get to market easily. In his diary,[2] Wood records the names of his staff in the early 1940s: Ndala (the cook); Daud and Ulanda (house-boys who cleaned the house, washed and ironed clothes, and performed other house duties); Masi and Jim (watchmen); Lamek, Mangani, Alibi and Patrick (gardeners); and Vincent and Lorens (who looked after machines, pumps etc. – probably on a part-time basis). The trustworthiness and reliability of staff was a very important aspect of life because the climate was often enervating, and everyone had to be self-sufficient and to rely on their own resources. Wood evidently had a good rapport with his staff, and he was able to leave them in charge when he went away. When he returned, he always found both house and garden in good order. On one occasion, typical of many, he wrote: '[Arrived home,] where all the staff welcomed me. Found the house and everything spotlessly clean, and the orchids in good shape etc. Daud has done marvellously well . . .' Wood's house was simply furnished; he had lots of books (mainly on natural history and gardening, and good literature), cabinets for his butterflies and shells, and a desk where he wrote letters. He had a radio (battery operated) and a gramophone (which was wound up with a handle). Over the mantelpiece was a lovely picture of a fish eagle, one of the majestic birds of prey that lived along the shores of the lake. There was no mains supply for electricity, so at night lighting was supplied by kerosene bush lights and Aladdin pressure lamps; later on, however, there was a generator (which was far from reliable). Rainwater was stored in a large tank, hot water was provided by a 44-gallon 'Rhodesian boiler' wood-fired water heater, and meals were cooked on a wood-fired iron range. During the dry season, the garden was watered by a pump bringing water from the lake – when the pump worked. These domestic services were standard throughout all the more remote places in Nyasaland at the time.

The warm, nutrient-rich waters of the lake support large populations of many species of fish, and some of them are an important source of food for local people. Most fishing during Wood's time was from dugout canoes which ventured no more than a few kilometres from the shore. Many people were involved in fishing, and it was not uncommon to see several dozen canoes bobbing about on the blue twinkling waters of the lake. In Chinyanja, the local language, fish are called *chambo* – a general name that includes a number of species that grow to a sufficient size (8–12 inches long) to provide a good meal. Another group of smaller fish, collectively known as *nsipe*, are much smaller; they are usually laid out to dry in the sun on wooden racks made from sticks, and then transported to distant markets for sale as 'relish'. The survival and wealth of many villages depended on a regular supply of *chambo* and *nsipe* from the lake. *Chambo* is one of the most edible freshwater fish in Africa; it is tasty, moist and succulent, and undoubtedly Wood would have eaten *chambo* several times each week.

Although Lake Nyasa is justifiably well known for its *chambo*, it is better known outside Nyasaland for the large number of species that live in the lake. The first fish from the lake to be given scientific names were collected by John Kirk during the Livingstone expedition.[3] In the following years, more fish were collected and sent to the British Museum

162

of Natural History. But it was Wood's collection of fish from Domira Bay in 1920 (see Chapter 4) which gave the first hint of the great biodiversity of species in the lake and which was the catalyst for many subsequent studies and collections. At present, there are about 250 described species.[4] Most of these belong the family *Cichlidae* (and are commonly known as 'cichlids'). There are more species of cichlids in Lake Nyasa than in any other lake in the world, even more than in the larger Lake Tanganyika to the north, where the coastline is similarly broken up into beaches and headlands. Many of the cichlids are beautifully coloured and patterned and, not surprisingly, some of them are now well-known as aquarium fish in many parts of the world. Ever since this marvellous diversity of fish became known, biologists have investigated how – and why – so many species have been able to evolve within a single lake. There are many reasons, but the basic one is that the lakeshore is made up of numerous small sandy beaches, each of which is separated from adjacent beaches by rocky headlands. There are also many substrates and depths close to the shore. Studies have shown that the cichlids are rather sedentary, and do not swim from beach to beach, or headland to headland. This is the perfect situation for speciation (the process whereby a species may evolve into a different species), because each species had been isolated in its own little habitat. Once a 'new species' is reproductively isolated from a closely related 'species' (by whatever means this is achieved), the new species may evolve even further from its ancestor in colour, pattern, habitat requirements and biology so that it will remain reproductively isolated even though it comes in contact with other related species. This process, repeated many times in different parts of the lake and over millions of years, has resulted in the great diversity of species found today. One of the magical experiences of Lake Nyasa, one that Wood must have experienced hundreds of times, is to float on the surface of the warm freshwater wearing a pair of goggles, and to look down through the crystal clear water at the myriad brilliantly coloured fish swimming and darting over the golden sand and among the dark rocks.

In Wood's day, boats were an integral part of life at the lake, as they still are today. They ranged in size from dugout canoes to Arab dhows (a legacy of the slave trade) and various forms of motorised boats. There were small boats with outboard motors, commercial fishing trawlers, and larger passenger and freight ships. In the late 1940s and early 1950s there were still some of the old mission boats, such as the *Chauncy Maples*, in service. Nyasaland Railways operated several passenger and freight vessels, including the *Ilala* (a namesake of several previous boats) and the *Mpasa*.[5] Perhaps most surprising of all was the *Herman von Wissman*, the former German gunboat that was captured by HMS *Guendolen* in 1914 (see Chapter 4). After the war, she was renamed *King George* and subsequently, when she became a freighter, she was given the name of *Malonda*. All these boats called in at Monkey Bay from time to time. Wood occasionally took a passenger boat from Monkey Bay to Chipoka on the western shore, but he was usually seasick and did not enjoy this method of travel. At times, when the road south from Monkey Bay was impassable, the only way to reach Blantyre and the south from Monkey Bay was to go by boat to Chipoka and then take the train that ran between Salima, a few miles north of Chipoka, to Blantyre. During the two periods of time when he was employed by Nyasaland Railways at Monkey Bay, he was closely involved in the maintenance and building of ships. There is no record of his duties, but he probably looked after staff affairs and the stores. In this capacity, he was indirectly involved with the building of the *Vipya*, which was to be the flagship of the lake steamers at the end of the Second World War.

The *Vipya* was designed and built in Glasgow and shipped in sections to Nyasaland, where she was reassembled in Monkey Bay, a process reminiscent of the first *Ilala* brought out by David Livingstone one hundred years previously. By lake standards, the *Vipya* was a big vessel – 373 tons, 140 feet long, 27 feet wide, and with a draught of about 8 feet. She was able to carry 315 passengers, 35 crew and 100 tons of cargo.[6] By the end of 1945 she had been reassembled, and by May 1946 she had completed her trials and went into service. Her normal passage was from Monkey Bay to Chipoka, and then northwards to Nkata Bay, Mbamba Bay (on the eastern side of the lake in Tanganyika), Florence Bay and Karonga, where she turned around and came southwards again. On 26 July 1946 she set off on her fourth voyage. All went well to begin with, but she encountered very bad weather after leaving Nkata Bay on her way to Mbamba Bay. Sudden squalls, high winds and rough water are characteristic of the lake and a very real danger to boats of all sizes. The captain, Commander Keith Farquharson (formerly of the

Royal Navy), decided to shelter at Mbamba Bay for the night in the hope that the storm would have abated by morning. On the next morning, 30 July, the weather was better but not perfect, and the *Vipya* set off again for Florence Bay. Exactly what happened at about 10.30 a.m., when the *Vipya* was 10–15 miles from Florence Bay, is still uncertain. The combination of strong tides, wind, large waves, sudden strong squalls, and the cargo hold being slightly open ready for unloading caused the *Vipya* to heel violently to starboard, turn over on her side and sink within about 90 seconds. It was thought that there were about 190 people on board (exact records were unavailable); of these, there were only 49 survivors, all Africans except for the Indian engineer. The subsequent Court of Enquiry heard evidence from many people involved with the building and running of the ship, and concluded that the captain should not have set out from Mbamba Bay and that he did not ensure that the decks and holds were made sufficiently waterproof when water splashed over the decks.[7] Other observers had commented previously that the *Vipya* seemed to be top-heavy and hence could be unstable in bad weather. A subsequent appeal produced all sorts of conflicting evidence, some of which showed that the weather at Mbamba Bay had not been as bad as previously thought and that the decision to sail across the lake was reasonable. The outcome was that the captain was exonerated. At the time, there were many rumours as to the cause of the accident, and despite the enquiry and appeal, it is doubtful whether the real reason why such a modern ship suddenly turned over and sank will ever be known.[8] For people at Monkey Bay, as involved as they had been with the building of the ship, the news was particularly shocking. Wood's comment in his diary on 2 August echoes what must have been felt by all Nyasalanders: 'Just heard the dreadful news that *Vipya* has struck a rock near Florence Bay and sunk with all hands. So ends the ship I helped to build on only her third voyage. And I was going up on her in October to Tanganyika.' A few days later he wrote: 'Hear that it was no rock but she just turned turtle in heavy seas 7 m[iles] from Florence Bay thro' being top heavy.'[9]

The importance of fish to the nutrition of lakeside people, and to their economy, had long been recognised by the colonial government. In the late 1930s, the British government funded the 'Nutrition Survey'.

This survey collected a lot of data on the fish and fisheries of the lake,[10] and showed that several species of *Tilapia* were the most important commercial species. The war interrupted these studies, but they were resumed again between 1945 and 1947. In 1945 a young fisheries officer, Rosemary Lowe, was sent out from England; she spent two years on the lake, mostly in the region of Monkey Bay. Here she came to know Wood very well, and her reminiscences provide a wonderful glimpse into life at Monkey Bay at this time.[11] In the mornings, they bathed in the bay just outside Wood's house, and then returned to the *khonde* for toast and tea. Under the roof of Wood's house, there was a row of mud-daubed swallow's nests which the swallows patched up each year. Wood used to talk to the swallows as they flew backwards and forwards or sat on the special perches placed across the *khonde* near their nests. They also watched the Cordon-Bleus and other small birds at the birdbath just outside the *khonde*. They then went to collect the fish from the gill nets that had been set overnight. The fish were brought back to the house, where Rosemary Lowe dissected them in Wood's bathroom to obtain their stomach contents for analysis. She recalls that they had many interests in common and enjoyed many discussions together:

> We discussed many things: Tanganyika and the United Nations, what a mess Britain was making in the colonies, why some butterflies were mimics, whether speciation was partly due to predation [one of the theories in those days], how many individuals were necessary to maintain a population . . . He had a wonderful library – a lot of books about Africa . . . he was apt to carry on about things he did not care for . . . the 'Ecological Society' – 'The one Society which I will not join. Why complicated statistics to prove everyday facts!' He was incredibly knowledgeable about many things, and was glad to have someone to talk to. It was wonderful walking with him in the bush – he talked about the birds and butterflies and things like that . . . It was fun staying there![12]

At other times, Lowe lived on the eastern lakeshore of the peninsula, close to Dally's Hotel:

We used to try and listen to the news in the evening on the radio. There was a cockroach which lived in the tuning bar of the radio, so the cockroach had to be pushed across while one was tuning . . . The start of the rains was an exciting time . . . all these scorpions and things used to come out of the ground. I had to walk from my home to the hotel, about 100 yards I think. I used to go along the beach because there used to be crocodiles. Once there was a lion which jumped on the milk-boy. I had elephants in the garden, and a leopard took the watchman's dog. You know, that's what it was like then. So I used to walk along the edge of the lake with my Tilley lamp to have my dinner and then crawl back. It was great fun![13]

Life in Nyasaland in those days was full of amusing incidents:

There were a lot of old timers living along the lake. A lot of beautiful people – really splendid people, like Arthur Dent, FitzMorris the doctor, and Peter Selous who was the DC. There were all sorts of silly things that happened. When all the askaris were coming home after the war, they were to be paid off and [so] they sent £1,000 in 5/- notes in a metal box in a lorry with the key hole facing up . . . so it filled up with water. So one of the first jobs was who to trust to dry out the notes . . . We had to go into Fort Johnston to get petrol. It was all in 4 gallon tins. . . . No petrol pumps. No electricity up there in those days. . . . Everything came in 4 gallon tins, in 'debbies'. We used them for collecting bath water, we used them for everything. I don't know how life would have carried on without debbies! Of course it was a fact of life when I went there – you couldn't get anything. I did a roaring trade eventually in old motor car tyres for people to take to pieces to make fishing nets. I used to buy fish with salt and money. We had to buy food – chickens and eggs and things. And of course you always floated the eggs to see how fresh they were. I don't remember getting many vegetables. I think the Greeks used to bring supplies [for Dally's Hotel] in the fish lorries.[14]

Travelling was not easy, especially if you did not own your own car. Rosemary Lowe recalled how she used to travel to Blantyre:

In the dry season it would be all right . . . and of course when it rained, it was awful. The way I used to travel was on the fish lorries; that was the standard way of going down [to Blantyre]. Yanakis [who owned a fisheries on the lake] used to run the big fish lorry, and used to export [dried] fish right down to Rhodesia. If you were lucky, you got the seat alongside the driver. If not, you sat on top. It was a lovely journey. There was no bridge over [the Shire River at Liwonde]. You always got there in the middle of the night, and had to wake up the ferryman. A lot of these places were [full of] tsetse fly in those days so travelling at night was better.[15]

At the conclusion of Rosemary Lowe's surveys, much of the work on fish and fisheries was taken over by the Joint Fisheries Research Organisation of Northern Rhodesia and Nyasaland (JFRO). Several young fisheries officers, including Peter Jackson, Geoffrey Fryer, Derrick Iles and Derek Harding, worked on the lake in the late 1940s and 1950s; they were a dynamic team and their work produced more evidence of the wonderful diversity of fish in the lake, and of the lake's importance as a place for studying the mechanisms of evolution in aquatic environments. The JFRO team lived at the north of the lake, and only visited the southern part of the lake occasionally. Peter Jackson met Wood on a number of occasions, and he remembers that they used to shoot fish together on the lake:

I had a Westley Richards .318 hunting rifle, the famous 'Accelerated Express', a very popular sporting rifle in its day. Geoff Fryer was working on the Mbuna, the little brightly coloured rock-dwelling cichlids which I am sure you know – they later became famous aquarium fish. The heavy 250-grain bullet of the .318 could and did stun a little fish in 3–4 feet of water. Rodney and I rowed about in his little boat hunting mbuna in the rocky coast around Cape Maclear. His skill as a rifleman was such as to let him do the shooting while I did the rowing. The .318 with its big bullet is a very potent weapon to me and most people, but Rodney treated it as if it was an airgun. He would whip it to his shoulder, aim and fire all in the same second while I was much too slow in my handling of it. Most of these old-time hunters were like that.[16]

There was a lot of wildlife on the peninsula in the 1940s and 1950s. Leopards lived in the rocky hills surrounding Monkey Bay and Cape Maclear. On many occasions, they got into the chicken and duck enclosures, killing a number of birds each time. In his diary, Wood recalls what happened one night:

> At 0330 a leopard jumped wire into fowl's enclosure and dashed about crashing at the wire trying to get out; much growling and noise; staff came on scene with lanterns and spears; I got out with revolver and electric torch; beast was in thicket in middle of the large enclosure and I could see nothing; sent staff to make a noise at far end of enclosure while I hid behind door of nearest hut at near end; leopard made terrific dash at wire in middle, out of my view, tore it from its supports and escaped; just as well![17]

More serious was when a leopard entered a hut and seized a young boy; his father, who was sleeping nearby, managed to drive the leopard away, but not before the boy suffered severe laceration of his thigh and minor wounds elsewhere. After Wood had given first aid, the boy had to be taken to Fort Johnston (some forty miles away) for proper treatment. Occasionally lions visited the peninsula, and their pugmarks were seen on the road near Cape Maclear and around the local houses. Sometimes there was an elephant, but by this time they were rare and seldom seen. Baboons and monkeys were a continual menace because they raided the vegetable garden and the fruit trees, and sometimes a hippo came up from the lake at night to feed in the garden. Squirrels nested in the roof of Wood's house, making a lot of mess and damaging the curtains. Snakes were common most of the time although their numbers varied from year to year. The year 1947 was a 'good' year for snakes: Wood had to kill a poisonous or mildly poisonous snake every few days in his garden but he did not kill non-poisonous snakes such as egg-eating snakes and pythons. When he was clearing some ground belonging to Nyasaland Railways, he found fifteen puff-adders in six weeks.[18] On another occasion, the cook found a three-foot-long puff-adder curled up on the kitchen floor.

The lake and the major rivers of Nyasaland are home to crocodiles and hippopotamuses. In the early days of the Protectorate, both species were extremely numerous. The first European explorers commented frequently about the large numbers of hippos in the Shire River, and the danger they posed to the small boats. Likewise, crocodiles were feared because their stealth, ugliness and air of malevolence have always been symbols of something extremely evil to humans. They are primarily fish-eaters, but they will catch and eat any warm-blooded mammal that enters the water or comes close to the shore if they are able to catch it. Many Nyasalanders have been killed and eaten by crocodiles. Sir Harry Johnston commented in 1897:

> . . . men, baboons, lions, leopards, antelopes of all kinds approaching the edge of the water are liable to be seized and pulled under by the crocodile . . . many [local people] lose their lives every year as victims of the crocodile. As a rule, the crocodile never attacks human beings when there are a number of them together in the water. It is only when a man or a woman is alone that the crocodile makes his rush . . . At Fort Johnston, on the Upper Shire near Lake Nyasa, the crocodiles would rush up to the very bank and seize people heedlessly standing near the water's edge. Several of our Indian soldiers were killed in this way until the river bank was guarded by a palisade. The crocodile seldom eats its victim immediately it has been killed by drowning. It prefers to stow it away in some crevice or hiding place under the water until it is partially decomposed.[19]

Perhaps the most graphic description from the beginning of the twentieth century about the horror of crocodiles in Nyasaland was given by R. C. F. Maugham. Based on his own experience, he wrote:

> We will suppose then that the time is early morning or a little before sunset, which are well known to be the hours of greatest danger, during which, if the river banks be low, no person is safe within five or six feet of the water's edge. A woman probably with a small child secured to her back by a length of calico, proceeds to the river, a large clay pot in one hand and a small gourd calabash attached to a short bamboo stick in the other. She does not look at the stream; she is wholly unconscious and careless of the fact, the possibility of which does not even occur to her, that a few yards away, motionless, just beneath the surface, a great

dim form and two cruel eyes are attentively watching her every movement.

Intent upon filling her jar, and forgetful to everything else, she
stretches over the water and begins to ladle a few calabashes into the
water-pot behind her. Little by little she has waded in ankle-deep to
reach the clearer water a yard or two from the edge. Suddenly, and in
an instant, there is a terrific irresistible rush, a wild piteous scream, a
heavy splash, followed by a commotion in the deep water such as might
be made by some large body moving rapidly beneath the surface. Far
out, a moment later, a human hand is thrown up for an instant, then
disappears for ever. The water flows calm again; there is silence; a large
partly-filled water pot stands on the river bank, and a small calabash
attached to a short bamboo stick floats slowly down the stream. That
is all. The entire heart-shaking fatality has probably occupied from five
to ten seconds . . .[20]

The only way to control the numbers of crocodiles was to shoot
them. The government licensed a number of professional crocodile
hunters, whose job was to shoot as many crocodiles as possible. Crocodile
hunting was a commercial business because the belly skin makes very
good leather for shoes, bags and suitcases. One of these hunters was
Paul Potous, who recorded his life and adventures with crocodiles on
the lake in his book *No Tears for the Crocodile*.[21] Crocodiles are extremely
wary and alert, even though they seem to be slow, sleepy and lethargic
when resting on a sandbank beside the water. Potous used to travel to
known crocodile sandbanks in his canoe or catamaran that was powered
by an outboard motor. He and his African assistants usually set off just
after dusk to the place where they planned to hunt that night. They had
to be careful that they did not pass too close to hippos, which could
easily overturn the canoe if annoyed. When they got close to the hunting
grounds, the engine was turned off and they slowly and quietly paddled
towards their quarry. Potous sat in the front of the canoe with his rifle
across his knees and a powerful searchlight close at hand. At a suitable
spot, the searchlight was turned on. The eyes of a crocodile shine red
in the light and it may be possible to make out the rest of the head
resting just above the water surface. When a crocodile was sighted, the
canoe was moved closer and one of the assistants came forward; he

took the searchlight from Potous, keeping the beam on the crocodile all the time. If the canoe drifted close enough to the crocodile, Potous could then attempt to shoot it. For the crocodile to be killed immediately, the bullet had to penetrate the very small brain enclosed within the immensely thick bones of the skull. Immediately after the shot, there would be a great commotion as the crocodile thrashed around, and then, if killed, it would turn over with its belly floating uppermost. If it was merely wounded, the hunters would quickly follow it and attempt a second shot. Each dead crocodile was tied to a marker buoy so it could be located the following morning. These hunts went on for several hours, before the hunters returned to their camp for the night. In the morning, each crocodile was pulled behind the canoe to the camp – crocodiles are so heavy that it was only possible to pull one at a time – where it was skinned and the belly skin liberally rubbed with salt as a preservative. Crocodile shooting is an extremely dangerous occupation: it must take place at night, there are mosquitoes, and there is always the chance that a hippo or a wounded crocodile will overturn the canoe or that the thrashing crocodile will knock one of the hunters into the water.

There were considerable numbers of crocodiles on the peninsula in Wood's time, though there were probably no crocodiles in Monkey Bay itself because of the disturbance caused by people and boats. Wood sometimes went to some rocks near Mpembe Marsh in the late afternoon and early evening, where he could watch the many crocodiles that lived there. But crocodiles did not seem to interest him as much as mammals, birds, butterflies and gardens, and he seldom wrote about them in his diary.

Besides these larger and mostly unwelcome animals were many smaller species which were a constant delight to Wood, especially the birds. There are about 320 species of birds recorded from the Monkey Bay peninsula, and about 50% of these are considered to be 'common'.[22] Many of them would have lived in, or close to, Wood's gardens at Monkey Bay and Cape Maclear. From his *khonde*, Wood was able to watch the seasonal changes of birdlife. The weavers, bulbuls, waxbills, sunbirds, flycatchers and fish eagles were resident all the time; some species, such as some of the kites, cuckoos and kingfishers, were inter-African migrants that lived on the peninsula for only a few months each year according to the season; others, such as some of the warblers, sandpipers, swallows and swifts, were migrants from Europe that visited only during the European winter. Wood loved to attract birds close to his house; the bird tables were visited by scores of small seed-eating birds – weavers, mannikins, sparrows and finches – and the birdbaths were enjoyed by many species, from the small finches to large birds of prey.

In front of the *khonde* at Cape Maclear was a big baobab tree. One of the species that nested in a hole in the baobab was the Pearl-spotted Owlet. Wood was fascinated by these birds because of the 'false face' which they have on the back of the head. When it is sitting on a branch looking forward, the observer sees its light golden eyes, speckled facial feathers and whitish eye stripes. If the head is turned so the owlet is looking over its shoulder, the observer sees what appears to be two heavily lidded dark brown eyes surrounded by a white facial disc, and a strong dark beak. This second 'face' is due to the remarkable pattern of the feathers on the back of the head which mimics the pattern of the real eyes, beak and facial disc on the front of the head. As the head is turned forwards and backwards, the 'face' changes, providing a most uncanny effect to an observer and the illusion of being watched all the time![23] After the Pearl-spotted Owlets had reared their young, the hole was taken over by a pair of Grey Hornbills. These birds have a most unusual method of nesting: the male hornbill builds a mud wall at the entrance of the nesting hole so that the incubating female is sealed safely inside. There is a small entrance hole through which the male feeds the female (and the young after they are hatched), and when the time has come for the young to fly, the wall is broken down and the female and young are released. By this time, the female is in poor shape because she has been unable to fly and her feathers look extremely dishevelled. Many other species besides these two also nested in Wood's garden, including bulbuls, flycatchers, sunbirds and shrikes.

Bushbabies lived in the trees around the garden. These delightful little animals, which weigh only about 200 grams, are related to monkeys. They are nocturnal, and have huge forward-looking eyes, large ears, soft fluffy hair, a long tail, and long, delicate fingers. They are extremely agile, jumping from branch to branch with amazing speed. One of them lived in the roof of Wood's house and used to come down to be fed with a banana in the morning and at dusk. Another one lived for a time in the kitchen and played all over the table and shelves. Otters were frequently seen playing and hunting in the water near Otter Point. Bush Squirrels lived in the trees and scampered around the garden. These are beautiful animals to watch, and often quite amusing, as Wood noted:

> 10.vi.53. Delightful comedy: Gabar Goshawk bathing at great leisure in the groundlevel birdbath; one of the squirrels (Paraxerus cepapi) of family living in a hole in the big Tamarind tree in front of house, got more and more excited and chattered and cursed Goskawk with fur fluffed out and tail over back; when squirrel approached bath, Goshawk raised wings over back and hissed at squirrel who retired to nearby bush overhanging bath. Squirrel very excited and finally Goshawk gave it best and flew into small 'timbiri' (Combretum sp.) on side of bath whereupon squirrel literally screaming with rage and excitement dashed up tree and routed Goshawk.[24]

Fish Eagles are one of the very special birds of the lakeshore. These magnificent large birds have a white head, chest and tail, chestnut shoulders and belly, and black wings. They live in monogamous pairs, and were a common sight in Wood's time as they glided above the lake and along the shorelines looking for fish. Their eerie haunting cry is one of the most memorable sounds of life at the lake, a cry that is immensely nostalgic to anyone who has lived on the lakeshore. Fish eagles hunt by patrolling above the water until fish are sighted; they then swoop downwards towards the fish and fly slowly close to the surface. While still flying with slow, measured flaps of their huge wings, the eagle rapidly dips its feet into the water and quickly grasps a fish with its sharp, pointed talons. Burdened by the extra weight, the fish eagle rises slowly upwards and flies to a favourite perch in a tree, where the fish is eaten. Wood must have been very fond of fish eagles: he had a picture of them in his living room. Occasionally fish eagles drop their prey, and on one occasion Wood benefited from this:

> 9.ix.55. Thanks to a sea-eagle, I had a fine big 'sungwa' fish for lunch. Mangani [one of the staff] saw one feeding by the reeds, drove it off, and bagged the fish which he sold to me! The eagle had only just torn away the gullet and part of the head; sides were perfect, fish over 12' long. Shows the power of these wonderful birds.[25]

Other animals that were numerous were bats. Although Wood never collected bats on the peninsula, he must have seen many of them at dusk as they left their roosts. Most numerous were Mastiff or Free-tailed Bats, which roost in colonies, often in roofs of houses. Although most bats are extremely beneficial to humans, Mastiff Bats can be a nuisance where there is a large colony. They have a rather strong, pungent odour, and they produce large amounts of rather smelly drop-pings. In 1957, while living at Cape Maclear, Wood found the smell from the roof of the 'Big House' so overpowering that he had to do something about it. He collected 25 debbies (four-gallon tin cans) of droppings from above the ceilings and used them as manure on the vegetable garden and the bananas. He managed to kill a number of bats, but since there were an estimated 4,000 to 5,000 of them in the roof, many would have returned the following night.[26]

Local rainfall, or lack of it, is a very important feature in the day-to-day life of Nyasaland. When there is too little, crop yields are low and food is scarce; if there is too much, crops are inundated and beaten to the ground, and there may be floods. Both situations occur with fright-ening frequency. The amount of rain in each wet season and the frequency and amount of rain in each storm are of crucial importance to the whole annual cycle of planting and harvesting. For their very survival, subsistence farmers rely on the amount of food that can be harvested and stored each year. Little is ever recorded when the rainfall is 'normal', but when there is too little or too much, it is noted in diaries and news-papers. The annual rainfall at Fort Johnston, not far from Monkey Bay, ranges from about 415 mm to 1,300 mm, with an average of 860 mm.[27] Most of the rain (90%) falls during the five months between November and March. However, rain may occur in any month of the year, but outside these months, it is light and extremely unpredictable. Even in the so-called wet season, there is huge variation between years. Analysis of the records for 19 years shows that in February (usually the wettest month of the year), the average rainfall is about 230 mm, although the actual rainfall can vary between 90 mm and 440 mm, the maximum being five times the minimum. Variation in the other months of the wet season is even greater than in February. It is hardly surprising that rainfall dictates the whole tempo of life in this part of the world.

Everyone in Nyasaland is thankful when the rains begin after the long, hot and desiccating dry season; but too much rain and cloud become wearisome and depressing, and after four or five months (provided the harvest is good), everyone is pleased when the rains stop, the sun shines, and the paths and roads are no longer waterlogged.

Wood kept his own records at Cape Maclear for a few years; during the years 1950 and 1956, annual rainfall varied from 583 mm to 1130 mm, with an average of 812 mm[28] – very similar to that at Fort Johnston. The maximum within any one month during the wet season was 530 mm (20.9 inches) in February 1955. Perhaps the most extraordinary rainfall event was on 5 April 1956, when Cyclone Edith passed over the Mozambique coast into Nyasaland; Cape Maclear received 125 mm (4.9 inches) of rain in a single storm! Days of extremely heavy rain are recorded periodically in Nyasaland, and usually cause extensive flooding and much damage to local crops and houses. Exceptional storms in the 1940s and 1950s occurred at Zomba in December 1946 when 700 mm of rain fell in 36 hours, and at Nkotakota in February 1957 when 570 mm fell in 24 hours.[29] The changing fortunes of the wet season are summed up by Wood's comments during the wet season of 1947–1948:

22.x.47. First rain of season, at 1730 hrs, following violent blustery wind and heavy stormclouds. A sharp thundery type shower of which most fell over bay, but not enough round my house to wet ground thoroughly. Lake very rough.

12.xi.47. First real rainfall of season. Last night at 20.45 hrs some distant thunder and later drizzly rain started. Went on all night and continued today until about 12.00 hrs; sky heavily clouded.

1.xii.47. On night of 30.xi. Heavier rain fell all night and local people planted maize today.

3.xii.47. Steady medium rain all night and light drizzle all day today [same for 4.xii.]

5.xii.47. Very heavy rain all night.

18.xii.47. First heavy thunderstorm last night. Violent thunder and lightening [sic] but little wind. Much heavy rain for a long period of early morning, tailing off at 09.00 hrs. Lisumbi stream came down and broke open sandbar mouth to lake.

5.i.48. Heavy rain for last two nights. Stream down in torrents. No damage on my place or Bouguet's as I had both well bunded and terraced but front of hotel, where nothing had been done, a deplorable mess of wreckage of flower beds and gullying. I had warned them a year ago of what would happen, but no notice taken. A sad sight.

Heavy storms and overcast skies continued throughout January and February 1948. On many days, dense clouds of nkhungu (or kunzu) flies were driven southwards across the lake. Streams and small rivers, some of them without water in the dry season, became rushing torrents. The soil was saturated, the country was a mass of tall green grass, and many roads were impassable. By March, much of the country was flooded. Wood's diary continues:

11.iii.48. Most violent rain for long period in night; all country around flooded including lower half of my vegetable garden.

18.iii.48. Short thunder storms continue daily. Lake rising very rapidly. Has flooded much of vegetable garden through water seeping up so this a.m. I had hundreds of small fish caught and put into the water to try to control mosquitoes breeding therein. We are oiling the rain-pools everywhere but the place is full of Anopheles mosquitoes all over the bay area.

21.iii.48. Thunderstorms continue daily and constant clouds of nimbus in all quarters; ground saturated and oozing all along base of hills.

29.iii.48. Rains easing off now; none since 24th tho' looks threatening today.

30.iii.48. Thundering all round throughout night with intervals of rain which continued till midday. Remaining overcast.

1.vi.48. Today lake level is equal to last year's high (1947). Rains still continue daily.

30.iv.48. Measured lake level surface 8' below chiselled line on rock at E. side of bay. This mark is maximum high level 1937 wave-wash line, actual level being about 4' below at dead calm. Water level still rising about 1' a week but probably just on end of rise for this year now.[30]

Another aspect of rainfall that interested Wood, as indicated in his diary, was the changing level of the water in the lake. Ever since the earliest Europeans saw the lake, it has been known that the level of the lake is not constant. Since then, water gauges to monitor water level and water flow have been placed in various locations along the lake and in the Shire River. Historical records have also provided information on presumed water levels in the past.[31] Lake level can be analysed for three different time spans: monthly, annually, and very long term. The monthly variation in level is 100–250 mm in the dry season, and 400–500 mm in the wet season; these monthly fluctuations are governed by local changes in rainfall and drainage, and although interesting, are not of great significance. In contrast, annual changes in water level may be quite significant and can affect the livelihood of people living close to the lake. The annual rise and fall in lake level is 900–1200 mm/year (3–4 feet), and can be as high as 1500 mm (6 feet) in a very wet year.[32] The fluctuations are due to the variations in water inflow into the lake during the wet and dry seasons. The wet season occurs from about December to April, although in some years it begins earlier and in others it ends later, and the dry season extends from about April to November. The rain in the early part of the wet season has little effect on the water level of the lake; most of this rain soaks into the parched soils, and only when the soils have become saturated, or nearly saturated, in February and March, does water begin to drain into the rivers and then into the lake. As a consequence, there is a 2–3-month delay after the rains begin before the lake level starts to rise. The rise continues into the early dry season until all the water stored in the streams, swamps and soils of the catchment area has drained into the lake. It takes several months for all this water to leave the lake, and hence the lowest lake level does not occur until the end of the dry season in about December. In general, the highest level of the lake is between March and June, and the lowest level is between November and January.

The seasonal changes in lake level are determined by many factors, such as the amount and pattern of rainfall, evaporation from the lake surface, the volume of water that leaves the lake along the Shire River, and the changing pattern of sandbanks at the southern end of the lake.[33] The lake is very deep and contains a vast amount of water (see Prologue), so the annual fluctuations account for only a very small percentage of its volume. The lake is bordered by the Rift Valley escarpments, especially on the western and north-eastern sides, so that the catchment area, although steep, is not particularly large. In total, the catchment area (including the lake itself) is about 48,850 square miles, covering most of

Malaŵi immediately to the west of the lake and a lesser area of hill country in Tanzania and Mozambique to the east of the lake. The lake itself covers 11,430 square miles of this area, giving a land catchment to lake ratio of about 3:1.[34] Many large rivers flow from the escarpments into the lake. The annual rainfall varies greatly at different places on the lake; it is highest (about 2,000 mm/year) in the north near Karonga and at Nkotakota, and lowest in the south (about 1,000 mm/year) near Monkey Bay.[35] Although these amounts of rainfall are not excessive by tropical standards, rainfall is restricted to only 4–5 months of the year.

Evaporation of water from the surface of the lake reduces the amount of water in the lake and the amount that is available to flow down the Shire River. The annual evaporation has been calculated to be 1,900–2,000 mm per year,[36] equivalent to the annual rainfall in the wettest part of the catchment area. However, evaporation occurs on only about one quarter of the catchment area (that is, on the lake itself), and hence the volume of water available to flow down the Shire (or to contribute to rises in water level in the lake) is equivalent to about three-quarters of the annual rainfall.

The amount of water leaving the lake depends upon the water inflow from the rivers, which in turn depends on the annual rainfall. If the lake is high, and is followed by several years of high rainfall, the Shire River becomes wide and swollen and may flood further downstream. Alternatively, if the lake level is relatively low and followed by several years of low rainfall, the Shire River may not flow at all. The historical records, and the measurements from the stations on the lake and the Shire, suggest that the following changes have occurred since 1850. When Livingstone and Kirk sailed up the lower Shire, their boats were frequently beached on sand banks, suggesting that water level in the lake was low (but not excessively low) at the time; however, once on the upper Shire, they were able to sail all the way from Matope to the lake. The level probably declined between the 1860s and the time when detailed records of lake level began in 1896 (the year in which the Protectorate of Nyasaland was proclaimed). In 1896, the annual maximum level (AML) was about 1,546 ft above mean sea level. Between 1896 and 1915, the level dropped by 8 feet to 1,538 feet when almost all flow in the upper Shire stopped and the upper reaches of the river were blocked by sandbanks,[37] some of which became overgrown with rushes and grasses. This is how Wood must have seen the river when he first arrived in Nyasaland. From 1915 to 1937, the lake level gradually rose, and by 1937 the AML was about 1,556 ft, some 18 feet above its low 21 years previously.[38] In the following years until 1960, the AML oscillated between 1,552 and 1,556 feet, and has not at any time declined to the low levels that were present in the early years of the twentieth century. In past eras, the lake level was much higher; the Stone Age tools and artefacts found by Wood at Mwenilondo, near Karonga, indicate that the lake level was about 130 feet higher then than in the nineteenth and twentieth centuries.

The level in the lake is dependent on a very delicate balance between rainfall, evaporation, and the outflow into the Shire River. The level at any time is not totally dependent on the conditions in any one year, but is affected by the conditions in previous years. Each of these variables are independent of each other although, in the long term, they are closely interrelated. Water inflow into the lake is dependent on annual rainfall, but also on the amount of that rainfall that eventually flows into the rivers. One of the major changes which affect lake level, and which is continuing to an even greater extent today than it was during Wood's lifetime, is the reduction of the number of trees and the density of the vegetation on the catchment area of the lake. This has three major effects: first, the lack of vegetation allows the rain to beat upon the soil rather than falling gently and percolating slowly into the soil; second, more water runs over the bare soil and into the rivers, causing soil erosion; and third, water is not 'held' within the trunks, branches and leaves of the vegetation, and consequently, evapotranspiration from leaves is reduced. As a result, the water cycle, the formation of clouds, and the amount and pattern of rainfall are adversely affected. The gradual increase in water level of the lake during the twentieth century may be due, in part, to the changing landscape of the countryside around the lake.

One of most significant changes in transport after the Second World War was the introduction of regular air services. The weekly flying boat service from Durban to Poole in southern England (see Chapter 7) was resumed in 1946. The journey took six days, with overnight stops at Mozambique, Kampala (on Lake Victoria), Khartoum (on the Nile),

Cairo and Augusta (Sicily). In addition, the flying boats made short stops at Lourenco Marques and Beira on the first day; Lindi, Dar-es-Salaam, Mombasa and Kisumu on the second day; Laropi and Malakal on the third day; Wadi Halfa and Luxor on the fourth day; and Marseilles on the sixth day.[39] There was considerable governmental pressure for the flying boat service to include Nyasaland. Cape Maclear was chosen as the most suitable place for flying boats to land, because the bay was sheltered and not hemmed in with hills, and a hotel had been built there a few years previously. There was no proper road to Cape Maclear, so passengers joining the flying boats had to travel by boat from Monkey Bay, or fly to the little airstrip that had been made at Cape Maclear close to the beach. In October 1949, the Solent Flying Boats of British Overseas Airways Corporation (BOAC) which flew between Johannesburg and Southampton scheduled an overnight stop at Cape Maclear: a north-bound Solent arrived on Thursdays and departed on Fridays, and a southbound Solent arrived on Sundays and departed on Mondays.[40] The trip to Southampton took three and a half days and cost £146 single and £262-16-0d return, inclusive of overnight stops. However, the Solent service was doomed almost from the start because many new aircraft, which could fly faster and more economically from land-based airports, were being brought into service. The flying boat service lasted only until October 1950. As a result, the Cape Maclear Hotel lost much of its custom and was abandoned. A few years later, in the 1960s, the hotel was dismantled and the stones were shipped across to Senga Bay near Salima, and incorporated into a new hotel there. After this short period in the limelight, Cape Maclear reverted again to a quiet, peaceful beach.

Wood travelled on one of the flying boats in 1946. With his interests in the countryside, it is certain that he enjoyed gazing through the windows at the scenery below. In those days, planes flew at a fairly low altitude and comparatively slowly. Wood recorded his thoughts when he flew from Blantyre to Dar-es-Salaam in October 1946:

23.x.46. Left Blantyre by air for Beira (in D.H. Rapide). I being the only passenger, we deviated from the main flight route and flew down the swamps below Marromeu, at 100–200 ft altitude over huge herds of buffalo, waterbuck etc. a sight I have not seen for years. Weather magnificent. Reached sea some 60 mi north of Beira, whence along coastline at 100 ft altitude all way to Beira, where arrived at 09.40 hrs. To Savoy Hotel for food, and at 12.00 to flying boat (Short Sunderland, Empire Class, 'Coorong') leaving Beira about 13.30 for Mozambique (Lumbo). Weather magnificent all way. Smooth trip. Lunch served aboard. Arrived Lumbo ca 16.30 hrs where tea and night at excellent B.O.A.C Hotel do Lumbo.

24.x.46. Off before dawn by flare-way. Sunrise indescribably lovely, with incredible colours on cloud floor below us (we at 6,000ft); breakfast aboard. Touched down at Lindi for fuelling. Weather deteriorating, much intermittent cloud right down to sea level. Bumpy in cloud masses

25 x. 46 Arrived Dar-es Salaam at 09.30 hrs . . . [There is no record of where the night was spent].[41]

In those days, pilots wrote a flight information sheet at regular intervals which was circulated to the passengers, informing them about the flight. On this flight, Wood kept one of the information sheets.[42] It recorded that at 11.35 GMT (13.35 local time), the airspeed of the flying boat was 180 mph, the ground speed was 140 mph and the height was 7,000 feet. At the time, they were passing close to the mouth of the Zambesi River, and were 30 minutes ahead of schedule. Passengers on this flight were also informed that they would be passing Quelimane on the port side in 25 minutes, and that the estimated time of arrival at Mozambique was 14.25 GMT, 16.25 local time, where the night would be spent at the BOAC resthouse at Lumbo.

The year 1947 was an important one for Wood because of three events which determined his life and travels for the remainder of his life. The first was that his old friend Arthur Westrop returned to Nyasaland to settle permanently on Magombwa. In the years since Westrop purchased Magombwa (see Chapter 5), he had lived in Malaya, where he worked for Malayan Fertilisers. When the Japanese invaded Malaya, he moved to Singapore, but was unable to escape when the Japanese overran the island in 1941. For the remainder of the war, he was a prisoner of war in Changi and Sime Road. On release, and after finalising his affairs in Malaya, he settled down as a tea planter and became a leading figure in the Boy Scouts of Nyasaland. He and Leslie, his wife, resumed their friendship with Wood, who visited them on many occasions.

The second event was that Wood realised that he could return to the Seychelles, where he could live very cheaply and indulge in his love of beachcombing beside the sea. And third, in early 1947, he met Sir Alfred Beit. Sir Alfred was the son of Sir Otto Beit and nephew of Alfred Beit,[43] both of whom were very influential figures in the development of southern Africa, especially Rhodesia. Alfred Beit was a friend and partner of Cecil Rhodes, and was particularly interested in the development of railways, which would enhance trade and development within southern Africa. It was Alfred Beit who financed the bridge over the Limpopo River at the border between South Africa and Southern Rhodesia,[44] and which is now known as Beitbridge. He was a wealthy man, but also a generous one, and he liked to support and endow worthy causes. Sir Otto Beit was similarly involved in southern African affairs, and was Director of the British South Africa Company, and of Rhodesia Railways; he was Trustee of the Rhodes Trust and the Beit Railway Trust, and he founded the Beit Memorial Fellowship for Medical Research.[45] Sir Alfred Beit was some 14 years younger than Wood, and was educated at Eton. He was for some time a Member of Parliament in Great Britain (1931–1945), and he also maintained his family's interest in Africa; like his father, he was Trustee of the Beit Railway Trust and the Beit Fellowship for Medical Research.[46] Sir Alfred and Lady Beit were great patrons of the arts; they supported many artists and collected paintings and sculptures from different parts of the world. The Beits, like Wood, were captivated by the beauty and tranquillity of the Cape Maclear peninsula. In 1947, Sir Alfred purchased some land at Cape Maclear between Otter Point and the site of the old Livingstonia Mission, where he planned to build a house and plant a large garden. When Sir Alfred and Lady Beit first visited Wood at Monkey Bay, Wood recorded that they 'came over for a chat on gardens, shell collecting, etc. [Sir Alfred] had previously sent me a present of Hutchinson's '*A Botanist in southern Africa*'. A most kindly and charming man.'[47] A few months later, when they met again, Wood recalled that 'Sir A. and Lady Beit arrived and spent a day on my place discussing schemes for my helping them at Cape Maclear . . . I to return from Seychelles in April 1949 and start the outdoor work and as "Factor". I agreed to do so, if

I keep fit. They to build my cottage there as soon as possible.' Thus began a delightful friendship which lasted for ten or so years until Sir Alfred sold the house at Cape Maclear. Wood always commented on how much he enjoyed the company of Sir Alfred and Lady Beit and the stimulating conversations that they had together. One can imagine the mellow good-hearted repartee that must have take place between the 'Old Etonian' and the 'Old Harrovian'! Sir Alfred and Lady Beit did not live permanently at Cape Maclear; they came for holidays for a week or a few weeks at a time, and for the remainder of the time, Wood had this corner of Cape Maclear all to himself.[48]

The house that Sir Alfred built was in the Dutch Cape style. Peter Jackson, who visited the house in the 1950s, recalls that the walls of the house were 'covered with paintings in every medium – portraits and bright and gay watercolours, summer gardens, flower-bowls and that sort of thing; carvings, sculptures and pottery all around'.[49] The land around was terraced with bunds to reduce erosion (bunding was one of Wood's favourite activities) and many fruit trees were planted. The house was designed by architects ('preserve me from professional architects and contractors' Wood lamented while the house was being built!); generators were used to supply electricity, and electric pumps pumped water from the lake for the garden. There was a motor launch called *Bonnie* and a sailing dinghy, *Kalulu* (meaning 'hare' in Chinyanja) for travelling to Monkey Bay because the road leading to Cape Maclear was impassable for much of the year; and there was a V8 Safari car for good weather. Wood sold his house in Monkey Bay to Nyasaland Railways, and gradually moved his possessions to Cape Maclear. To begin with he lived in the Cape Maclear Hotel, which was still flourishing at that time, but he got fed up with hotel life because of the 'noise, banging doors, howling winds and incessant cold draughts'. So he moved into the stewards' quarters which had been completed at Sir Alfred's house until his own cottage was finished. Life was simple and satisfying: the climate was, for the most part, warm and calm, and the garden was always full of flowers and birds. Gardening, feeding the birds, letter writing, swimming and goggling in the bay, and supervising the 'Big House' and the staff occupied each and every day. Supplies had to be collected from Monkey Bay, mostly in *Bonnie*, but also in the Safari car when the road was passable. Visitors came to stay, providing good conversation and a break from the usual routine.[50]

During the years at Cape Maclear, Wood planted and tended Sir Alfred's garden. It contained masses of flowering shrubs and trees, and in time was almost like a luxuriant jungle. Many of the plants were common and would be well known to anyone who has had a garden in tropical Africa: frangipanis of many colours, oleanders, lantana, tecomas, bougainvilleas, quisqualis, antigonon, tabebuia, cassias, thunbergia, alla-mandas, flame trees and bauhinias. The early wet season was a particu-larly lovely time, when most of these plants burst into flower, and many of them continued to flower throughout the wet season. The flowers attracted a lot of birds, and the density of the bushes and trees provided many nesting places for them.

During these years, Wood developed a passion for orchids. Sir Alfred was an orchid enthusiast too, and together they amassed a huge collec-tion, said to have been one of the best in central Africa. Sir Alfred sent orchids from South Africa, Wood brought others from East Africa and the Seychelles, and some came from England and America. On their arrival, Wood planted the orchids in pots or in Osmanda compost, attached them to logs or bark, or tied them in the axils of trees. Most orchid specialists are both knowledgeable and fanatical about orchids, and Wood was no exception: he was fascinated by the complexity and colour of the flowers. On most days, he kept records of which orchids were in flower and of their colours. On 22 February 1954, he wrote[51]: 'One of the *Ansellia* I brought from Mombasa opened a spike of 10 flowers today; the prettiest I have ever seen, yellow with heavy blotches and large spots of very dark chocolate brown. I think *Ansellia* classifi-cation is in chaos, it may be *nilotica*, under which name I shall index it.' And on 8 March, he recorded: '*Laelia gouldiana* opened first flower today, a deep exquisite tint of rose-purple. *Dendrolobium* has two buds on the stem. Received . . . a nice specimen of *Voluta junonia* from Gulf of Mexico'.[52] In 1954–1955, there were 50–60 species of orchids in the garden, all of them festooned on the branches of trees. During these years, Wood kept meticulous records of the beginning and end of flow-ering of each species. Some species flowered all year, others for only a month or two, but in every month there were many species in flower at the same time.

In about 1946, Wood began to take a great interest in seashells, which satisfied his collecting instincts as well as allowing him to enjoy beachcombing and swimming in the sea. He had collected a few shells in the early 1930s on the coast of Natal, but there were no further opportunities until 1946 when he visited the East African coast, and until after 1949 when he lived in the Seychelles (see Chapter 10). He became passionately interested in shells and was very knowledgeable about them, just as he was about butterflies. He was more interested in the large beautifully patterned species, especially cowries and cone shells, and not especially interested in the smaller, less colourful species (those that were often scientifically more interesting). Wood preferred to collect live animals and prepare the shells himself. Occasionally he bought shells from fishermen, or in a bazaar, but only when they were in perfect condition. Hence his collection consisted of only a limited number of perfect specimens of each species. He never collected land shells or freshwater shells. In 1946, he visited the islands near Dar-es-Salaam purely for the purpose of collecting shells. Most of his time was spent on Mafia, Chole and Juani islands. It was an adventurous time, sailing on dhows (which frequently had to be rowed when there was no wind) and sleeping rough on beaches and in little guesthouses. All his collecting had be done at low tide, and in many places he was disappointed to find a lot of dead coral – and hence very few shells. Wood recorded that Chole and Juani islands were depressing places: there were many decaying and ruined buildings, very little food (because of drought), and the local 'freshwater' was brackish and useless for shaving. The water did not make good tea, so he usually drank coconut milk.[53] Nevertheless, he returned home with about 300 shells, which were carefully catalogued and placed in cabinets.

Among the most trying aspects of living upcountry were the bad state of the roads and the difficulties of keeping machinery in working order. In the dry season, travelling by road was possible, but was dusty and bumpy. The plumes of dust of an approaching car could be seen long before the car itself was visible. The road into Cape Maclear was particularly bad; it was narrow and twisted, crossed many stream beds and had numerous short but very steep hills. Wood made himself 'honorary road supervisor' of this road, since he did not trust the Public Works Department to look after it adequately. The road was impassable after the first few heavy storms of the wet season, and remained closed (or very treacherous) until the end of the rains. It usually took several months to rebuild the drifts, fill in the potholes and washouts, and clear the ditches on the roadsides. Mechanics from the railways had to be brought to Cape Maclear to help with the road work. Cars, boats and water pumps were constantly having mechanical failures; spares took weeks to arrive, and when the water pumps failed in the dry season, all the water for the garden had to be carried in buckets and cans from the lake. By 1955, Wood was fed up: 'Really, this incessant breaking down of all machinery is fast driving me crackers . . . Thank God I am not a rich man with all the necessary worries and possessions they pile up!'[54]

Although Monkey Bay and Cape Maclear were rather remote in the 1940s and 1950s, there was a constant flow of visitors, who came to enjoy the lovely beaches. Many of them called on Wood, and retained vivid recollections of him but also of what it was like to live in Nyasaland at the time. One of his visitors in the late 1940s was Dr Wright, a Government Medical Officer. Dr Wright and his wife visited Monkey Bay for a holiday:

Monkey Bay was a pretty curved bay with an island in the mouth of it. Apart from the hotel and two bungalows, there were no other buildings; no shops, no petrol station, no banks, no electricity, no telephone, no refrigeration, no hospital, no clinic, nothing; a weekly truck to and from Blantyre. The only exception to all that was . . . a generator at the hotel so it was comfortable and well appointed. Rodney Wood was very well known in Nyasaland . . . and I was fortunate to spend a day or two with [him] when on holiday at Monkey Bay. He was then living in a small bungalow while engaged in supervising the construction of a landing strip for planes nearby, a job for which he was ideally suited with his profound knowledge of Chinyanja and his long association with the country. In his bungalow he had a number of cases of insects and he told us that he was an acknowledged authority on the blood-sucking flies of central Africa, some of which he had discovered for the first time and which were named after him. He was of a lean and weathered appearance and burned brown by all his years under the sun. He was bursting with life and anxious to communicate his enthusiasm to everybody. It was a privilege to be in his company . . . I can remember sitting on his verandah at sundown while he threw down handfuls of maize for the birds, a habit he had had for years. They appeared from nowhere, dozens, even hundreds of them. It was a wonderful scene.

He was also very proud of his filing system. He had built up a gigantic card index which was housed in a huge wooden construction that he had had made for him by a local carpenter, and he used it as an encyclopedia. He claimed that it could answer almost any enquiry made of it, so my wife put it to the test by asking it how to make a Victoria sponge cake. Not surprisingly it failed, but Rodney was not at all put out. He extracted a blank card, and solemnly took down at her dictation the relevant details and filed it away for future reference.

I have often thought that it was a pity that he lived before the advent of the computer, which I feel he would have embraced with enthusiasm. Likewise I think he would have been a keen supporter of the Gaia hypothesis. The only thing is that his mind was already overstocked so I don't know how he would have had time for anything much else.[55]

A few years later, while Wood was at Cape Maclear, Don Baring-Gould and his family were living at Monkey Bay, renovating the hotel there. On several occasions they visited Wood, and Baring-Gould found the road from Monkey Bay to Cape Maclear especially memorable:

[It was] 12 miles, and with 32 very narrow bridges and in a couple of instances, two hills of 1 in 4 where one had to change into 1st gear or you slid down again – really hair-raising particularly if the road was wet! A notice at one them said: 'Stop – Select 1st gear Beware of back seat drivers!' Coming back from Cape Maclear one evening it bucketed with rain – we had to have seven attempts to get up one hill.[56]

The Baring-Goulds spent a weekend with Wood and were shown around Sir Alfred's garden:

The house was magnificently, fully equipped with all mod cons of that time – electricity, air conditioner (Rodney told us he did not believe in these modern contraptions). [There was] an absolutely superb garden at the side of the house with many tropical plants and fruits which were watered by an enormous rainer. The garden itself must have been the best part of two acres. My eldest daughter, who would have been about 5 at the time, was always invited to have a bathe under the rainer as, at that time, it was impossible to swim off the beach to the lake as there were far too many crocs. [He seemed] to hit it off really well with my daughter as he and she were dancing together for quite one half an hour to his gramophone. He said they got on well because they dressed alike he only in his shorts and she in her panties! I don't ever remember seeing him in a shirt. It was like being accompanied by a walking encyclopedia when we went for walks with him – he would tell us all the types of birds, whether male or female, their nesting habits, etc. It was fascinating![57]

It was Wood's huge knowledge of natural history and his infectious enthusiasm which drew people to him rather like metal filings to a magnet. Christopher Barrow and his parents used to visit Wood at Monkey Bay and at Cape Maclear. He remembered visiting Cape Maclear in the 1950s:

Rodney Wood had become Sir Alfred Beit's 'head gardener' . . . where he lived in a very open-air wood chalet near Sir Alfred's Cape Dutch villa. As you know he was an avid botanist and had incredible green fingers. The one drawback was that he had absolutely no idea of landscape gardening and succeeded in no time at all in creating a tropical jungle full of orchids, frangi-pani and such like. He loved to hold court on his khonde and I for one was fascinated and could sit for hours listening to him. . . . When he returned from his island in the Seychelles periodically, he would return probably illegally with new exotic plants and trees which he planted in Sir Alfred's garden. One such plant I knew was the pure white frangi-pani which had dark green leaves with rounded ends. It can now be found all over Malawi. He was a very wide spectrum naturalist with an encyclopedic memory of all the birds, animals, insects, fishes and plants. However, he was definitely a 'poacher turned game-keeper' as in his early days he was a renowned hunter with a number of record trophies to his name.[58]

Another visitor in the early 1950s was Brian Marsh who, at that time, was a young commercial crocodile hunter on Lake Nyasa. One of Marsh's hunting grounds was along the south-west arm of the peninsula, south of Cape Maclear. In those days, the coast was lined with reedbeds, and was mostly unpopulated except for a few small fishing villages. Brian Marsh used to visit Cape Maclear and made every attempt to see Rodney Wood:

Living alone at the Beit beach cottage, situated at the very top (north-east) end of this beach was one of Africa's most interesting and colourful characters, an elderly bachelor named Rodney Wood. I would go to any length to be invited to the beach cottage, and often was for lunch or dinner, so I could listen to Mr Wood's stories. To effect this end I would make my camp as close to the beach cottage as possible, knowing full well if Mr Wood knew I was there that I would be invited.

Mr Wood, as I always addressed him and so will continue to address him, was a Chinyanja linguist, and seemed to know the Chinyanja, Latin, and colloquial name of every animal, tree, shrub, flower, beetle, moth, butterfly and insect that occurred in Nyasaland. . . . His hobby in Nyasaland was walking in what was then the miles of virgin forest that occurred behind the cottage. I walked with him on occasion, and was treated to a running commentary which I found fascinating and extremely educational. On these walks, and in fact whenever he was out of the cottage, he always carried a butterfly net.

He was obviously a man who was completely satisfied with his own company and preferred to be alone, yet he always greeted me as though he was delighted that I had come. I liked and admired him tremendously, was awed by him really, and I remember him with great affection. . . . Although he must have been in his mid-sixties at the time I knew him, he certainly did not look it. He was rugged of countenance, his expression alert. He had sandy-coloured hair if I remember correctly, bright searching eyes . . . He was tall and slim, well proportioned, quick of movement, erect, spry, sunburnt, and I never saw him in anything but khaki shirt and shorts. He had a quiet voice and a measured way of talking. He always had plenty to say and talked constantly about things that interested me, but wasted few words on idle chatter.

When I lunched with Mr Wood, he would always select a book for me to read while he went for his afternoon nap, telling me to read it and not swap it for something else. His books generally were too highbrow for me then, but I became familiar with authors names which I have since read and greatly enjoyed. He introduced me to Paul Gallico with 'The Snow Goose', and when I had not finished reading it by the time I had to leave, allowed me to take it. He was very fond of classical music, and always had a record on the wind-up gramophone during dinner. He was a marvellous raconteur, and I could listen to his stories for as long as he was prepared to keep talking.

[Mr Wood] had a very fine double-barrelled .475 Nitro Express . . . I understood that he bought this when he became game ranger but confessed that he seldom used it preferring to shoot everything . . . with his bow and arrow. When I asked why he resigned from such a fascinating position, he replied that the powers that be had got fed up with him because of his reluctance to shoot crop-raiding animals – which was accepted government policy. Unless an elephant or buffalo or hippo was a danger to life, he would tell a complainant to build a fire in his dimba garden and beat a tin when the marauders came near – not to call him out to shoot everything that came near.

He was a keen antelope bow-hunter, using an 80-pound long bow and arrows which he manufactured himself from dowels he bought at the hardware store and metal heads which he cut from discarded strips of hoop-iron used as strapping around imported grain bags. On one occasion, he gave me an arrow and told me to shoot it down the beach. I had never used a bow before – let alone one with an 80-pound pull – and I sent the arrow all of 25 paces. He then took the bow and sent an arrow all but out of sight! Mr Wood averred that he almost never lost an animal wounded with an arrow . . .[59]

The idyllic life at Cape Maclear came to an end in about 1958 when Sir Alfred sold his house and Wood's employment as 'factor' came to an end. In spite of the joys of Cape Maclear, the constant problems of bad roads and mechanical breakdowns, the endless rain of the wet season, and his intolerance of cold were getting too much for him. His various sojourns on the Seychelles, which he was coming to love more and more, were happy breaks from 'work' at Cape Maclear, and he longed to spend more time on the islands. Over the years, Wood had visited the Westrops at Magombwa, and in 1956, perhaps knowing that Wood was shortly to leave Cape Maclear, they offered to build a small cottage for him next to their house – the house that Wood himself had built in the 1920s. It was an ideal situation for Wood, and he marked the occasion by presenting a book to Leslie Westrop in which he inscribed 'RCW to Leslie Westrop to celebrate a great decision'. As the cottage was being built, Wood made various visits to Magombwa, landscaped the garden around the cottage, and brought plants and orchids from Cape Maclear. Finally, in September 1958, he packed up the last of his possessions and took up residence at Magombwa. In his diary, he wrote: 'Grand to be away from CM at last and up here'.[60]

Chapter 9

MAGOMBWA AND LA ROCHELLE: 1958–1962

It took the best part of a year, on and off, for Wood to get all his plants, furniture, collections and other *katundu* from Monkey Bay to Magombwa. Often friends collected the odd piece of furniture or a few plants and took them to Blantyre, from where they found their way eventually to Magombwa. Within a few days of his arrival, Wood was out in the garden with Leslie Westrop placing orchids in trees and planting shrubs. The vegetation around the house had grown immensely since Wood left in 1928, and the 'dingle' in the valley below the house was now a well-grown forest.

The last years of Wood's life were spent alternately at Magombwa in Nyasaland, at La Rochelle in Southern Rhodesia, and on Cerf Island in the Seychelles (see Chapter 10). One might have thought that Wood, in his late 60s and early 70s, would have been content by then to remain in one place studying his birds and his collections, and enjoying the company of his many friends. But his nomadic inclinations had not diminished over the years. During 1958 to 1962, he made fourteen major journeys back and forth between Nyasaland, Southern Rhodesia and the Seychelles.[1] He travelled mainly by train and boat, and occasionally by air. On all his journeys, he travelled with lots of *katundu*, including plants, seeds and boxes of his precious shells.[2] Before each departure, he packed boxes to take with him or to be stored in his absence; on arrival, he had to unpack and get organised again. Every box or case was numbered, and its contents recorded.

Living at Magombwa, close to his old friends the Westrops, must have been like living in the past. By this time, the Westrops had lived at Magombwa for about ten years, and were well known in the area. Arthur Westrop was very active in local affairs: he was a scoutmaster and much involved with the Scout Association throughout the country, and he was also churchwarden of the church at Cholo. Wood's cottage had been built by Arthur and his son Richard (who was also Wood's godson); it was small and simple (like all of Wood's 'shacks'), with just a living room, a bedroom, a bathroom and a large *khonde*.[2] From the *khonde*, Wood could view the garden and the birds that came to his bird-baths and bird tables. Some enormous trees shaded much of the cottage, keeping it cool in the heat of the day, and providing a forest-like environment all around.

One of the old retainers, Jackson, looked after the cottage and cooked simple meals for Wood. A typical day began with breakfast on the *khonde*, watching the birds feeding and bathing in the coolness of the morning. Later on, Arthur, Leslie and Wood would go for a bird-watching walk through the plantations and forests of the estate, or along the Nswadzi River. Lunch was in Magombwa house, and afterwards Wood retired to his cottage to write letters, sort his stamps (apparently a new collecting craze for him), catalogue photographic slides, or give attention to his orchids, which were festooned on the trees around the cottage. Afternoon tea was at Magombwa, listening to records or talking to visitors. A sundowner and dinner ended the day. At night there was electricity until about 9.30 p.m., when the generator was turned off. When Leslie drove to Blantyre or Limbe, Wood accompanied her and spent the day shopping or dealing with his affairs. On one occasion, he visited the newly founded Nyasaland Museum, which was housed in the old Mandala resthouse – a place of nostalgia, he recalled, 'where I spent my first night in Blantyre 50 years ago'.[3]

Many interesting people visited Magombwa. In 1959, Digby Lewis arrived to sample blackflies in the Nswadzi River,[4] in much the same place where Wood had collected in the 1920s, and where he found the blackfly that was later named as *Simulium woodi*.[5] In the years since *S. woodi* was described in 1930, the species has been found also in Tanganyika, Kenya, and Northern Rhodesia,[5] and has been the subject of extensive research.[6] Blackflies are the vectors of a minute nematode worm that causes 'river blindness', or onchocerciasis, in humans. In the early days in the Cholo district, the disease was quite common. Much has been learnt about it since Wood collected *S. woodi* and other species of *Simulium* at Magombwa. The larvae live in fast freshwater streams (such as the Nswadzi and many others in the Cholo–Mlanje area), and the late larval forms and pupae attach themselves to the hard shells of freshwater crabs. When the adults emerge, they bite humans and other warm-blooded mammals and in this way transmit the nematode from one infected human to another. Surprisingly, Lewis (with Wood's help) found only one larva and one adult during two weeks of extensive searching (although he found larvae and adults of other species). Digby Lewis was surprised at the decline in numbers, and in 1960 he wrote about why he thought the decline had occurred:

> The Cholo area has undergone much deforestation in recent decades, largely for tea cultivation which had developed since 1928, long after tea-growing started in Mlanje. . . . In the Cholo area, the clearance of trees and bamboo and other undergrowth is said to have increased silting in some streams. Mr R. C. Wood informs me that in 1917 the area was covered with woodland composed of *Uapaca kirkiana* Muell. and *Brachystegia* spp. and the local climate was damper so that mosses and lichens grew thickly round Magombwa house outside which he collected the first spec-imen of *Simulium woodi* that year.[7]

Another visitor of great interest to Wood was Peter Hanney, who was the first curator at the Nyasaland Museum. Wood conceived the idea of such a museum in the 1920s, but at that time there was no money and no political will to establish a museum for natural history and ethnography. When, in the last years of the colonial period, a museum was established, Wood wistfully commented: '40 years too late.' Being the curator of a museum in Nyasaland is a job he would have loved, although undoubtedly he would have been frustrated by the red tape of administration and bureaucracy. Peter Hanney's particular interest was small mammals,[8] just as it was one of Wood's interests in 1914–1920; while curator, Hanney made a large collection of mammals for the Nyasaland Museum, but this was mainly after Wood had died in 1962.

Another visitor to Magombwa in these years was G. D. Hayes, who, like Wood, played a very significant role in conservation in Nyasaland.[9]

He probably met Wood in about 1925, when he first went to Nyasaland to work on a tobacco estate, and their paths would have overlapped until Wood left Nyasaland in 1931. Hayes, like Wood, hunted in his early days, but later became a staunch conservationist. In 1945 (when he returned to Nyasaland after the war), Hayes was appalled to see a letter in the *Nyasaland Times* entitled 'Game must go', in which the writer argued that conservationists were starry-eyed idealists and that there was no room in Nyasaland for agriculture and the larger game animals and hence the game must be eradicated.[10] Such a letter is reminiscent of letters of the 1920s, when the 'game–tsetse fly–sleeping sickness' debate was at its height (see Chapter 6). Hayes was determined that such views should be vigorously challenged, and hence he was instrumental in founding the Nyasaland Fauna Preservation Society (NFPS), whose ideals were based on those of the Fauna Preservation Society in London. The NFPS, over many years, has done valiant work in educating the public (including government servants) and by making submissions to the newly formed Department of Game, Fish and Tsetse Control. It is difficult, from this distance in time, to appreciate the uphill battle that was necessary to establish the game reserves and National Parks that now exist. Certainly they would not have come into existence without the dedication and enthusiasm of Hayes and the NFPS. One can well imagine how much Wood and Hayes had in common, and although Wood took no active part in post-war issues of conservation, it is likely that his views had a strong influence on Hayes. Thus, when Hayes visited Magombwa (and Cape Maclear in earlier days), they had a lot to talk about. Hayes acknowledged Wood's contribution to conservation when he wrote in 1978: 'Lengwe National Park probably owes its existence more to the late Rodney C. Wood . . . than to any other factor.'[11]

In 1959, Wood visited the Inyanga Highlands in Southern Rhodesia for the first time, to stay with Sir Stephen and Lady Courtauld. While he was at Cape Maclear, Wood had met Sir Stephen and Lady Courtauld, but how this meeting came about is not recorded. The Courtaulds had a beach cottage at Koko Bay, not far from Monkey Bay, and undoubtedly heard about Wood's interest in plants and gardening from their neighbours. Gardening is one of those English interests which has always transcended the established social boundaries of class and position.

Sir Stephen's family originated in France, near La Rochelle in Brittany, but as Huguenots, they fled to England in the late 1600s to avoid persecution by the Catholics. In England, they continued their trade as goldsmiths, later turning to silk, and later still to textiles and chemicals.[12] From these enterprises, the Courtaulds became very rich. Sir Stephen's brother, Samuel Courtauld, was the chairman of Courtauld Textiles for many years (1921–1946), although it seems that Sir Stephen was not involved with the company to the same extent. Sir Stephen, who was six years older than Wood,[13] and his Italian wife, Virginia, lived in England for much of their lives. In the 1930s they acquired the lease of Eltham Palace on the outskirts of London. For more than five hundred years, the palace had been a royal retreat; it had a magnificent medieval hall and was surrounded by fourteen acres of gardens. By the early twentieth century, the palace was delapidated and the Courtaulds spent a lot of money and time bringing it back to its former glory.[14] Sir Stephen was an art collector, and specialised in collecting Turner watercolours. He was also very keen on yachting, and he and his wife toured the world in their ocean-going motorised yacht. At the end of the war, the Courtaulds decided to leave Eltham Palace because they were 'sick of Europe' and wanted to live in a place where ' human nature [was] much more in the raw'.[15] So they bought an aeroplane and flew all over Africa (where Lady Courtauld had spent some of her childhood), looking for their ' Shangri-La' where they could begin a new life. After a lot of searching, they settled in the Penhalonga valley near Umtali, high up on the Inyanga Mountains of Southern Rhodesia. Here they built La Rochelle, designed to remind Sir Stephen of one of the old Courtauld homes at La Rochelle in France. It is an estate of 182 hectares on the forested hillsides of the Imbezi Valley. The house is large and spacious, all on the same level, with reception and dining rooms, and about six bedrooms; the most noticeable feature on the north side is a circular tower with a conical roof reminiscent of that on an old French chateau.[16] Outside there are guest chalets and 'The Fantasy', where Lady Courtauld had her own study. The house is surrounded by 16 hectares of gardens. Close at hand are the orchid houses, the greenhouses, the vegetable garden and a swimming pool. Further away is the Dell, an artificial lake surrounded by magnificent 'foxglove trees', and several special gardens devoted to aloes, palms and conifers. In places there are little streams and aqueducts, shaded by tangles of trees and shrubs, and thick with ferns and mosses. The formal garden immediately in front of the house is like a parkland with extensive lawns, rose gardens and beds of colourful annual and perennial flowers.

Wood stayed with the Courtaulds on four separate occasions between 1959 and 1961. It is perhaps not difficult to understand why the Courtaulds and Wood developed such a close relationship: their mutual interest in plants, orchids, gardens and shells were a strong basis for friendship, but also, no doubt, the Courtaulds enjoyed Wood's company and stimulating conversation and were appreciative of his immense knowledge of natural history. Lady Courtauld often called him 'Professor Timber', a delightful play on words which also expressed her appreciation of his abilities.

Travelling from Nyasaland to Umtali was not difficult in the late 1950s. The train left Limbe at 12.45 p.m. and arrived in Beira at 9.00 a.m. the following morning. Wood described the journey as very comfortable. The train descended from the highlands into the Lower Shire Valley, through Chiromo (close to where Wood lived from 1914 to 1920) and Port Herald, and southwards to the Zambesi River. By this time, the bridge had replaced the old ferry and passengers could stay in the same coach all the way from Limbe to Beira. Because the evening train did not leave Beira until 5.30 p.m., Wood spent the day in Beira, usually visiting friends. He arrived in Umtali at 7.15 the next morning, where the Courtaulds met him and drove him to La Rochelle.

Living at La Rochelle must have been a wonderful holiday for Wood – he described it as '[a] truly marvellous place'. The Courtaulds were very liberal-minded people and there was a constant stream of very interesting visitors of all nationalities and classes. There were musical evenings, and visiting dancers. Some idea of the people who visited 'La Rochelle' is recorded in Lady Courtauld's unusual 'visitor's book' – a huge window of plate glass on which visitors inscribed their signatures with a diamond stylus pen. It provides a fascinating glimpse at what life was like at La Rochelle. Wood's signature is near the middle of the window, and close by are the signatures of Prime Ministers, Ambassadors and authors, amongst others.[17] One can imagine that Wood revelled in the controversial and stimulating conversations that must have taken place every day when visitors were in residence. But he also had time for his other interests: he helped in the garden and gave advice about trees, organised the orchids which he brought from Cape Maclear and the Seychelles, walked with the Courtaulds in the garden and on the hills beyond, and spent a lot of time bird-watching and sorting his stamps.[18] His other main activity was cataloguing his shell collection. By this time in his life, he was thinking about what to do with his collection after his death. Lady Courtauld had told him that a new museum was to be established in Umtali and she persuaded him to donate the collection to the museum; in the meantime, she offered to display the collection at La Rochelle. Wood liked this idea, and even discussed the possibility with Reay Smithers (Director of the Museums of Rhodesia) that he might be involved in some way with the Umtali Museum.[19]

The Courtaulds took a very active part in the cultural life of Umtali. Because of their liberal views and their desire to improve the relationships between all races in the country, they became involved with the Rhodes Society, a multicultural group that was founded in 1958. The idea was that Africans, Indians and Europeans could have social evenings together, discuss topics of mutual interest, and participate in joint projects. The Courtaulds bought an old factory building which was converted into a hall and a Society dining room; later on, buildings for crafts, pottery and other activities were added.[20] Wood often used to call in there to talk to anybody who happened to be around; he clearly enjoyed listening to the many varied opinions of the members. During the two years or so when Wood stayed from time to time at La Rochelle, membership rose from 35 to nearly 400. The concept of such a society at that time, when white imperialism was the norm, was very novel and undoubtedly did a great deal to foster trust and friendship between the races. For a while, Wood stayed in a room at the building belonging to the Society while cataloguing its library, but he found the traffic noises so annoying and upsetting that he did not stay for long – he much preferred the quietness and peace of La Rochelle. Gwynneth Ireland, secretary of the Rhodes Society at the time, had an abiding memory of Wood when he was just over 70 years of age:

Rodney was so very interesting that it would be impossible to forget him altogether. He wore his then silver hair cropped very short – just a fuzz over his skull which made his ears more prominent (he said he resented

199

time spent brushing his hair). His skin was clear and very little wrinkled, but the character lines around his eyes and mouth had deepened considerably and his face was very much leaner. His eyes were a clear and sharp grey-blue and he generally had an expression of benign amusement. His frame was spare and deeply tanned. He still had enormous energy and his habitual walk was an easy loping stride.

The only clothes that I ever saw him wear were a well-washed and faded khaki safari jacket and shorts, and a pair of brown leather sandals without socks. He didn't bother to dress formally for any of the Society functions but nevertheless he was always welcome . . . He was every bit as courteous and encouraging as a listener as he was a raconteur, and unless asked a direct question about his accomplishments did not volunteer much about himself. He told me his income was only £30 per month (remember this was in the 1960s when the average wage for a teacher was about £900 per annum) and he often undertook various tasks for the love of the work in hand and/or board and lodging, rather than financial return. He didn't despise money so much as he usually ignored it. I believe that most of his travels in his latter years were paid for by his many friends who found it a small price to pay for his interesting and informative company.

The Courtauld's estate at Penhalonga . . . had about eight small houses on the property to house their African staff, with a small homecraft hall to house the meetings of the local African Womens' Group to learn handicrafts and various activities. Rodney greatly enjoyed his visits there and captivated the women by his attention to what they had to contribute more so than what they had learned . . .

Rodney and the Courtaulds had much in common. They all possessed a great deal of energy, had travelled widely, loved books, music, art and the natural sciences (especially botany, lepidoptera, entomology, herpetology and conchology) and understood the need for careful recording in all branches. Had he lived in these times, Rodney would probably have been a prominent member of Green Peace.[21]

One of the Courtaulds' special projects was the establishment of an agricultural training school in the Tsonzo area, 35 miles from Umtali.[22] The school was named Kukwanisa, which, in the Shona language, roughly means 'being able to do things well'. They obtained the land from the government, paid for the building of the school's facilities, and financed an endowment trust to provide the capital for running the school. It took many years to complete the enterprise, and it was not until 1964 that the first 18 students began their studies. The farm was sufficiently large that, when fully operational, it was able to give a good all-round training in agriculture to about 120 students. Wood visited the Courtaulds at the time when the farm was being planned, and no doubt was able to give advice based on his past experiences as a planter and instructor in agriculture at Livingstonia. The Courtaulds took an immense interest and pride in the development of the school, and used to visit it almost every week. They also funded many buildings in Umtali: the theatre (now called the Courtauld Theatre); Queen's Hall (for concerts and other large functions); and the Tea Pavilion at the Show Ground. Many other organisations also benefited from the Courtaulds' generosity: the Bulawayo Theatre; St Michael's Church in Harare; the University College of Rhodesia; the National Gallery; the National Museum; the National Trust of Rhodesia; and the Rhodesian Academy of Music.[23] Their final bequest was to give La Rochelle to the National Trust of Rhodesia. Although both Sir Stephen and Lady Courtauld were contro-versial people, with views that were often contrary to the establishment of the day, they embraced Africa and Africans throughout the twenty years that they lived in Southern Rhodesia.

Wood's last years saw many changes in Nyasaland. Although he was not interested in politics and government (although very willing to criticise bureaucracy), he must have been aware of the emerging political conscience of local people. In 1953, Nyasaland became part of the Federation of Rhodesia and Nyasaland.[24] The Federation was composed of three member states – Northern Rhodesia, Southern Rhodesia and Nyasaland – and was primarily an economic and political arrangement. It may have brought some economic benefit to Nyasaland, but indige-nous politicians did not approve of their country being dominated by the larger and richer Rhodesias. To them, nothing short of total inde-pendence from Britain and from the Federation was acceptable. In the late 1950s, the Nyasaland African Congress was formed to lobby for independence, and in 1958, Dr Hastings Banda returned to Nyasaland as its leader. Dr Banda had left Nyasaland in 1915 to look for work in

South Africa. After he had done various menial jobs (which he did not enjoy), an American mission arranged for his schooling and university studies in America. In 1937 he qualified as a medical doctor, and then went to Britain, where he was in general practice until 1953. When Nyasaland was forced into the Federation, he decided that he could no longer live in Britain, which, in his view, had betrayed the interests and aspirations of Nyasaland. So he went to Ghana, where he lived until returning to Nyasaland in 1958 after an absence of 43 years. Political activity increased after Dr Banda's return, and with his demands for independence. On 3 March 1959 was a turning point for the political future of Nyasaland: the government arrested about seventy political activists at Nkata Bay, and about forty demonstrators were killed while trying to release the activists. Dr Banda himself was arrested, the Congress was banned, and a state of emergency was declared. These events intensified political unrest rather than diminishing it. The outcome, eventually, was the release of Dr Banda from prison in April 1960, constitutional talks in London, internal self-government, and the breakup of the Federation in 1963. Full independence was granted on 6 July 1964, and Nyasaland became the Republic of Malaŵi.[25] Wood missed all the events in 1959 because he was either in the Seychelles or at La Rochelle', but he would have been witness to many of the events of 1961. He did not live to see the breakup of the Federation, nor independence. Probably he would have been rather bemused by such a rapid change in so short a time and, like all old men, would have preferred to think of Nyasaland as he knew it in the 'old days'.

More importantly, for Wood, the changes in the environment that he had seen throughout his fifty years in Nyasaland would have been cause for greater concern. Most of these changes were the result of an increase in the number of humans, better medicine and living standards, and the expansion of agriculture. When Wood first arrived in Nyasaland in 1912, there were about 969,000 Africans,[26] and by 1960 the number had increased nearly threefold to 2,800,000.[27] The increase in numbers was accompanied by an increase in the area of land used for subsistence farming, for commercial crops such as tea and tobacco, and for plantations of exotic timber. The extent of natural vegetation declined as a result of all these activities. As an example, in the 1920s, the road which Wood cycled or drove along from Cholo to Blantyre was bordered at that time, almost as far as the eye could see, by *Brachystegia* forest. The valleys and Cholo Mountain were thickly clad with montane forests and there were still a few elephants living there. There were no tea plantations and no farms, and hardly any Africans lived there because it was so cold at some times of the year. The vegetation was self-sustaining because of the high evapotranspiration and rainfall, and the density of trees, shrubs and herbs provided a moist microclimate which encouraged the growth of many mosses and lichens, which were a noticeable feature of Magombwa in the 1920s. By 1960, most of this country had been converted to tea plantations and agricultural fields, and natural vegetation was to be found only in small, isolated patches. Similar changes in the environment occurred throughout the country to a greater or lesser extent: clearance for agriculture, plantations and urban development, and the constant demand for wood for building and cooking, drastically reduced tree cover. Most people do not notice these changes because they happen slowly, and it not easy for one generation of humans to realise what it was like when their grandparents were alive. But the fact of the matter is that the natural environment of Nyasaland in 1960 was very different to what it was in 1912 when Wood first arrived (and it is even more different at the beginning of the 21st century). It must have saddened him to see such large and rapid changes in the environment, and the irreparable loss of so many habitats and animals – all in the space of only fifty years.

These years must have been quite exhausting for Wood, as he was constantly travelling between the Seychelles, La Rochelle and Magombwa; however, travel was much easier and more comfortable than when he first came to Africa. At La Rochelle and Magombwa, he was treated as an honoured guest and well looked after. Although he loved his life in the Seychelles, Africa was still 'home' and he could not resist the lure of returning whenever possible. Arthur Westrop remembered that Wood was still as enthusiastic during his last visit to them as he had been in all the previous thirty-seven years of their friendship: 'During an evening walk along the Nswadzi, a large Charaxid butterfly flew overhead. "'That's *falvifasciatus*," Rodney exclaimed as he raced round the small tree to get a view of the point at which the butterfly had settled. "I never saw it on Magombwa before, it only occurs on the tops of mountains. I must get my net."[28] The net was found, and a piece of wooden curtain rod added for a handle. Wood spent many happy hours thereafter watching a bait of rotting bananas, brown sugar, rum and beer (a mixture which was most attractive to these magnificent butterflies) and hoping this special butterfly would return. Wood's carefulness as a collector had not diminished with age: most of his evenings were spent setting and cataloguing butterflies, and as Arthur Westrop wrote, 'his perfection in this direction was as remarkable as ever [and], as a ex-entomologist [myself,] a joy to behold'.[29]

On his last long sojourn in Nyasaland, and after some wonderful months collecting butterflies and living at Magombwa in 1961, Wood returned to the Seychelles in September 1961. When he left, he told the Westrops that he was uncertain when he would be able to return, probably because he did not know whether he would have enough money for the passage. Fate, however, intervened – he never returned to Nyasaland again.

A selection of Rodney Wood's butterflies in the Natural History Museum of Zimbabwe.

Top:
Family Nymphalidae.

Middle:
Family Pieridae.

Bottom:
Family Papilionidae.

Left:
Caricature of
Rodney Wood
by
R. G. Eldred,
Chief Medical
Officer,
Dedza,
Nyasaland,
1919.

Above left: Painting on porcelain of Jessie Francis Carrington, one of Rodney Wood's
aunts, about 1885.

Above right: Marcel Calais, Seychelles, 1998.

Left:
A fish preserving tank in wooden box, similar to that used by Rodney Wood to preserve fish from Lake Nyasa (Natural History Museum, London).

Above: Fish racks for drying fish on the shore of Lake Malawi near Cape Maclear, 1985.

Below: Lake Malawi near Cape Maclear, 1985.

Above: Tea-picking at Magombwa near Cholo, 1994.

Below: Mount Mlanje from near Zomba, 1985.

ve: Zomba Market, 1985.

w: Zomba Mountain from near Zomba, 1985.

Two views of the Central African Plateau of Malawi, where Rodney Wood conducted surveys for game and tsetse fly in the 1920s.

Above: Kasungu National Park, 1985.

Below: Nkotakota Game Reserve, 1993.

Lengwe National Park.

Top: Nyala at a waterhole, 1985.

Middle: Flooded track in the wet season, 1985.

Bottom: Dry grasses in the dry season, 1985.

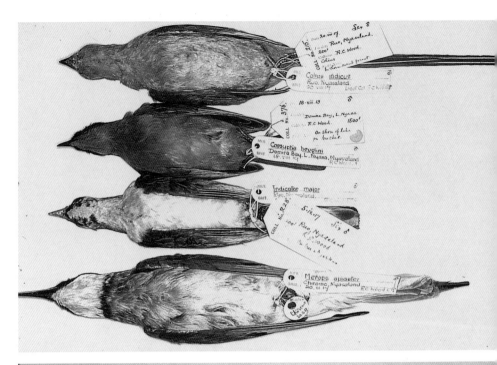

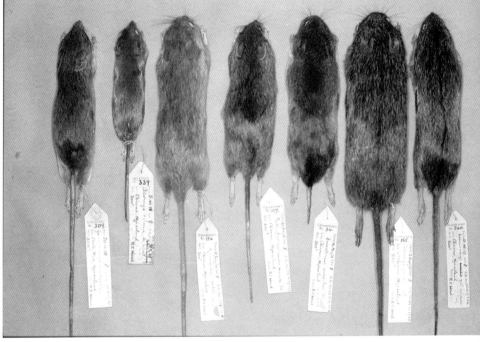

Rodney Wood's specimens.

Above: Birds in the Bird Room, Natural History Museum, Tring.

Below: Rodents in the Natural History Museum, London.

Above:
Bottles containing some of Rodney Wood's fish from Lake Nyasa collected in 1922, now in the Natural History Museum, London.

Left:
Wooden comb and paper-knife made in Nyasaland once owned by Rodney Wood.

Above: Ardvreck School, Crieff, Scotland where Rodney Wood was a student 1899 to 1903.

Below: La Rochelle, near Umtali, Southern Rhodesia, the home of Sir Stephen and Lady Courtauld. Rodney Wood visited on four occasions during the 1950s.

The Lower Shire Valley.

Above: The valley from the eastern escarpment of the Rift Valley; the Shire River is clearly visible in the middle distance.

Below: One of the rapids on the Shire River north of Chiromo.

Above: The house that Rodney Wood built for himself on Cerf Island, Seychelles.

Below: The view from Rodney Wood's house on Cerf Island towards the sea and one of the neighbouring islands, 1998.

Chapter 10

THE SEYCHELLES: 1921–1962

Wood's life in the Africa is interwoven with a totally different life in the Seychelles.[1] It is difficult to imagine two such different environments, but he embraced both of them. In Africa, he lived on a vast continent of savannas, forests, highlands, and freshwater lakes, and amid a great diversity of animal life; in the Seychelles, he lived on a small island surrounded by sea and coral reefs. Wood visited the Seychelles for the first time in 1921 on his way back to Britain after the First World War. Why he visited the Seychelles is not known – maybe he liked the thought of somewhere very different to the vast expanses of Africa; maybe he had been told about the beauty of the islands; or maybe it was just a whim which appealed to his nomadic instincts. Whatever the reason, he fell in love with the islands and they became his second home. The climate was warm, the pace of life was slow and leisurely, and the islands were far away from the troubles of the 'real' world. Wood visited the islands eleven times, staying between six and eighteen months on each occasion.[2] His first three visits were between 1921 and 1932. His next visit was in 1949, and from then until his death in 1962, he spent more time on the islands than in Africa.

The Seychelles are a group of granite and coral islands, perhaps the most beautiful tropical islands anywhere in the world. They are situated in the Indian Ocean, 1,600 kilometres east of the coast of East Africa, 1,500 kilometres north of Mauritius, and 2,800 kilometres southwest of India. Long ago, they were part of the great supercontinent of Panagaea, which was situated near Antarctica. Huge geological upheavals 200 million years ago split the supercontinent apart, and the landmasses now known as Africa, South America and India were formed and started to drift northwards. Another small fragment came to rest in the Indian Ocean, but only the granite peaks of the mountains are now above sea level. These peaks are the islands of the Seychelles. The best known of these are clustered close to one another, and include the islands of Mahé, Praslin, La Digue, Silhouette, Aride, Frigate, Cerf and St Anne. The largest is Mahé (150 square kilometres), which rises to a height of 914 metres, and the smallest is Round Island (*c.* 2 hectares). Over the millennia, coral reefs have developed in the shallow waters around each of the granite islands. Close by are two other islands (Bird Island and Denis Island), formed entirely of coral and rising only a few feet above sea level. Further away, and quite separate from the granite islands, are numerous other coral islands which are grouped in four separate clusters known as the Amirantes, Alphonse, Farquhar and Aldabra groups.[3] The most remote of these coral islands are the Aldabra group, about 1,000 kilometres west of Mahé, and well known as a World Heritage Site. Because of their geographical isolation from the big continents, most of the islands of the Seychelles have evolved endemic species of animals and plants – a feature which makes them a fascinating and unique place for biologists. Altogether, there are about 80 islands in the Seychelles, though this number increases to about 115 if isolated rocks as small as 0.5 hectares are considered as islands.

Because the islands are a long way from anywhere, the Seychelles were not 'discovered' until fairly recently. The first recorded evidence of the islands is given on a Portuguese map of 1501 which shows the outer islands of the Amirantes group. One hundred years later, in 1609, the island of Mahé was visited by Europeans for the first time. The islands were a haven for visiting ships because there was abundant fresh water, fish and fruits, and there were no human inhabitants.[4] The next visit was in 1740, when French ships from the Ile de France (now Mauritius) and Bourbon (now Reunion) spent several days at Mahé, where they re-provisioned with 33 giant tortoises and 600 coconuts.[5] The French annexed the islands to the Ile de France in 1756, and the first European settlers arrived between 1770 and 1785. Between 1770 and 1810, the islands were ruled by the French, but it was an ambiguous situation because ships of the British Navy attacked the islands on several occasions, and the French Governor was forced to capitulate. The French flag was lowered and the British flag was raised, but the French Governor still remained in charge. In 1810, the British occupied the Ile de France and Bourbon, and, by default, the Seychelles as well. These were uncertain times – French privateers and pirates were attacking British ships travelling between India and Britain, Napoleon had conquered Egypt, and Britain required a strong naval presence in the Indian Ocean to protect her interests. The Seychelles were an excellent base for the British Navy. But after the defeat of Napoleon in 1815, the islands declined in strategic value to Britain and 'entered into a period of long, peaceful semi-oblivion'.[6]

From about 1770, the islands have been populated by immigrants from Europe, Africa, Arabia and India. Some of the races have retained their original ethnicity, but intermarriage has produced a marvellous diversity of skin colour, hair texture and facial characteristics. Whatever their ancestry, the Seychellois people regard themselves as Creoles, a name which reflects the happy multiculturalism of the islands. Many languages, primarily English and French, are spoken, but the most wide-spread is Creole, a French patois with additions and modifications from English, Bantu and Hindi.[7] The first settlers introduced many new plants to the islands; at first, these were mainly vegetables, but later commercial plants such as coconuts, cotton, cinnamon, nutmeg, vanilla, cloves, peppers and patchouli were imported. The products of these plants have formed the backbone of the subsistence economy for many years. The islands have changed greatly since human settlement. Most of the original forest along the coast has been replaced by plantations, and now there are few places which are untouched by human activities. Nevertheless, the Seychelles still retain the magic and beauty of remote tropical islands where time stands still and the troubles of the rest of the world are of little consequence.

It is hardly surprising that Wood, on his first visit, was entranced by the beauty of the islands. From the deck of the ocean liner, moored a few kilometres offshore, he would have seen the huge granite boulders of Mahé rising out of the sea. Scattered among the boulders and hills were forests of luxuriant vegetation, and on the shoreline were bright golden sands bordered by coconut palms and rich dark green takamaka trees. There were few buildings to spoil the beauty of the shoreline. Probably there was a slight breeze wafting the delicious smell of spices across the waters of the bay. The water was deep blue and crystal clear, turning to turquoise and golden blue closer to the shore. Passengers were ferried in small boats from the ocean liners to the island; once in shallower water, they could have seen with unbelievable clarity myriad shoals of colourful fish and beautiful seaweeds swaying gently with the tide. For Wood, it was a totally different environment to the vastness and dryness of Africa, and one that he loved at first sight.

The climate of the Seychelles is warm and tropical with little daily or seasonal variation. During the day, the maximum temperature is 27–30°C and at night it is 24–25°C, a daily range of only three or four degrees regardless of the time of year.[8] Rain occurs in all months, but most of it falls between November and May. Typically there are about 2,300 mm (c. 90 inches) of rain, and about 190 rainy days, each year.[9] For most of the time the humidity is high (75–80%), and for humans, this can be rather enervating at some seasons. The warm moist climate is wonderful for plants: the Seychelles are richly covered with luxuriant vegetation and there are flowering plants in every month. The islands are the perfect example of a tropical paradise.

The only record of Wood's first visit to the islands in 1921–1922 concerns his purchase of land there. The purchases are recorded, in French, in huge leatherbound volumes in the Department of Legal Affairs in Mahé. It seems that Wood spent at least eight months in the Seychelles on his way to Britain after the war. At this time, he purchased four 'arpres' (approximately four acres) on Mahé, and 20 arpres on nearby Cerf Island. His second visit, of nearly a year, was in 1924–1925, when he was returning to Nyasaland from Canada and the United States. It was during this period that he started to keep a diary of interesting natural history observations and other noteworthy events. His third visit was in 1931–1932, after he had resigned as Game Warden of Nyasaland. His diary provides an interesting account of life in the islands at this time.

In 1924–1925, Wood spent much of his time at Maluwa, his property on Mahé. Maluwa was situated in the hills, not far from the capital, Victoria. In the fashion that was typical of all his homes in the years to come, he built his own 'shack' with the help of local workmen. There was no road to the property and all the building materials had to be carried up from the beach. During its construction, Wood lived in a tent (at least for some of the time). He particularly enjoyed terracing the garden and planting trees. He planted a lot of vanilla and cinnamon trees, cochin coconuts, Java coconuts, jamalacs, mangoes and bananas, and several sorts of citrus trees, and he lined all the paths with dense tussocks of lemon grass to reduce erosion. He made seed beds for the many seeds he had collected during his travels – pepper trees, date palms, *Cocos yatai* and *Washingtonia* palms from California, pomelo and rambutan from Singapore, and 'also two kinds of trees from Hawaii'. Flowering shrubs commonly found in gardens in the Seychelles and other parts of the tropics were planted around the house – bougainvillaea, hibiscus, allamanda, jasmine and thunbergia; and there were lots of orchids that he had imported from Penang. Wood was delighted at the rate of growth of his plants – cuttings were growing six inches each day. Very quickly Maluwa became a jungle.

Because of the remoteness of the islands, inhabitants had to make their own entertainment. It did not take Wood long to find his niche in the European community of Mahé; afternoon tea, or 'tiffin', with friends, and visitors 'dropping in' at Maluwa were part of the normal lifestyle. Very often, visits to friends were so enjoyable that they lasted longer than might have been intended; the Kenworthys – good friends at this time – jokingly said to Wood: 'We will call you Jorrocks for "where I dines I sleeps, and where I sleeps I has breakfast."'[10] All these social activities were very impromptu, simply because there was no means of rapid communication in those days. Wood became friendly with the Governor of the islands, and frequently went to tea and dinner at Government House. As in Nyasaland, there were many social circles and close-knit relationships, and little mixing between different circles. Other activities in Victoria in which Wood participated were a book club and a fencing group run by the Marist brothers; he enjoyed the rather unusual situation of fencing with a Marist brother dressed in a cassock.

Listening to music on his gramophone and on the radio was a favourite occupation for Wood. He preferred classical music and songs, and he enjoyed singing by himself and with small groups. In the evenings, he sometimes went down to the beach below Maluwa after supper and sang Hawaiian songs in the moonlight.[11] At other times he sang with small groups at concerts that were held either in Government House or in Carnegie Hall in Victoria. The concerts enabled residents to get together, exchange news and gossip, and have fun. Wood participated in many of these concerts, and usually sang several songs, either alone or with a group; at one concert in 1925, he was a member of the 'Orange and Blacks', a 'pierrot troupe' that sang and fooled around like clowns. For another concert he dressed up in a Robin Hood costume and sang 'The Song of the Bow' and 'The Yeoman of England'. It was at one of these concerts that Wood became well-known throughout the Seychelles: at the beginning of the concert, when everyone was seated, he appeared at the back of the hall dressed in his Red Indian ceremonial dress. Nadege Nageon, a very young girl at the time, recalled the lasting impression that Wood made on this occasion:

> When all were assembled, a tall lean and deeply tanned Englishman in full Red Indian regalia took three steps forward, aimed his arrow at the target (a beam set up at the far end of the hall) hitting the bull's eye with such force that the building shook on its foundations, setting off vibrations that lasted for 15 seconds. To those assembled it was a stellar performance to be remembered by one and all. RC had left his mark on the Seychelles . . .[12]

Although Wood liked his little hideaway at Maluwa, he much preferred to live on the edge of the sea. So, in mid-1925, he brought another property at Montfleuri on the coastline, not far from Maluwa. On his first night there, he listened to the gentle ripples of the sea on the shore, and henceforth called this property 'Rippling Waters'.[13] There was a small house which needed renovation because white ants were eating the foundations. He started to plant trees and shrubs and practised his archery on the beach at low tide, and many of his friends joined him for swimming parties. The numerous land crabs in the garden ate many of his new plants (a problem he did not have at Maluwa), but he solved the problem in his own special way – by shooting them with his bow and arrows. However, Rippling Waters did not satisfy him, and he sold it before returning to Nyasaland in 1925.

After several years at Magombwa and as Game Warden, Wood returned to Maluwa in February 1931 for a visit of 18 months. In his absence, the vegetation at Maluwa had grown wonderfully well, but despite the beauty of the location, Wood decided that it was not a place where he could put down roots. He was still looking for somewhere close to the seashore which was peaceful and tranquil. Eventually, after much procrastination, he realised that Cerf Island was the place he wanted to live. He had already visited Cerf in 1921 (when he bought some land there), and again in 1925, but it was not until 1931 that he finally came to the decision that Cerf Island was to become his home. On his arrival in the Seychelles in 1931, he stayed with Clement Nageon de Lestang, who was his notary as well as a friend; this was the beginning of a long and wonderful friendship with the Nageon family that was to span three generations. Within a few days, Wood had arranged to buy a small plot of land at the southern end of the island; however, this meant that he had to start building another shack and planting another garden.

Cerf Island (or Ile au Cerf) is one of a group of five islands – Cerf, St Anne, Moyenne, Round, and Long – which lie a few kilometres offshore from Mahé. The islands are separated from Mahé by the Cerf Passage, a channel of deeper water, and they are surrounded by coral reefs. Cerf Island is only 137 hectares in area, and the highest point on the island is 107 metres; only St Anne is slightly larger and higher (at 217 hectares in area and 250 metres high). From the sea, Cerf Island appears like an up-turned saucer; it is completely covered with deep green vegetation without any huge granite boulders or cliffs that are such a feature of Mahé. The coast is rocky, but interspersed with little sandy beaches, overhung in places by large takamaka trees whose large green shiny leaves glint in the sunlight and crackle in the breeze. There are no roads, no motor cars or motorcycles, or not even pedal bikes. A few little walking paths criss-cross the island, but usually residents go from one part of the island to another by boat around the coast. There are a few small streams which normally have water for most of the year, and at the north end a depression in the sandy soil becomes a little swamp after heavy rain. In Wood's time (and even now), the few houses that exist are set back from the beach, and are mostly invisible from the water. Cerf Island derives its name from one of the French frigates, *Le Cerf*, which visited the Seychelles on 1 November 1756 when Captain Corneille Nicholas Morphey, leader of the French Expedition, claimed the islands on behalf of France. On this day, there was an impressive ceremony in Mahé, when there was a nine-gun salute, the

French flag was raised, and the 'Stone of Possession' was placed in the harbour at Victoria. Unlike Mahé, Cerf Island has not participated in rapid development or colonisation by humans. In the late 1700s and during the 1800s, several families bought land on Cerf island and settled there. Their descendents still own and live on the island. Occasionally, a few outsiders (such as Wood) have bought land, but in general, the ownership of the island has remained within the founding families. The first of these was the Calais family from Brittany in France; later, the Delafontaine family arrived from Lausanne in Switzerland, and the Collie and Gardner families came from Scotland.

In the early days of the twentieth century (and earlier), Cerf Island was almost self-sufficient for its everyday needs. Parts of the island were cleared for farming. Small black and white cows grazed on the pastures under the coconut palms, and their milk was used for local consumption and to make butter. Some milk was taken across to Victoria for sale. There were chickens and ducks, and a few pigs. Coconuts were harvested to make coconut oil for cooking, and the husk of the coconuts was crushed to make meal for the pigs and chickens. Beans, tomatoes, pumpkins, gourds, aubergines, spinach, and chillies grew well in the sandy soil; papayas were common, but the soil was not rich enough for bananas. Fish were plentiful and were caught using basket traps set out on the reef, and were a staple part of the diet. There was no electricity, telephones or postal services; water was provided by rainwater stored in tanks and from the few streams. Food was cooked on wood fires or kerosene stoves, and kerosene lamps provided light at night. Every family on Cerf had a boat (or maybe several boats) that they used for visiting other people on the island, for fishing, and for going to Victoria to collect supplies and mail. Most boats were rowing or sailing boats, although by the 1920s and 1930s outboard engines were used occasionally.

When Wood first lived on Cerf Island, life had not changed much from that of the previous century. Most of the supplies needed to build his first 'shack' in 1931 – wood, corrugated iron, cement, window frames, and nails – had to be purchased in Victoria and brought across in his sailing boat. He employed carpenters and builders to help with the building, but he did much of the hard work himself. In his diary, he commented that he had many blisters and 'chipped' feet, but this was just part of 'the normal process of hardening up!' While building was in progress, he hired a little hut to live in, as well as a cook and a houseboy to look after him. In time, the house was finished and furniture was moved from Maluwa. Nearby, Wood built a little 'laboratory' shed for the fish and shells that he collected on the reefs, and a boat house to shelter his new boat, *White Tern*. In the garden, there were birdbaths which attracted mynahs and doves, and bird tables with rice, bread and bananas. Two species attracted to the garden were particularly noteworthy. One was the small Seychelles Falcon, locally known as *katiti*, which was cinnamon and white in colour and boldly etched with black markings, rather like the European Kestrel. Katitis enjoyed bathing in the birdbaths; but sadly they were eating the geckos and nestling birds, so in the end, Wood had to get rid of them. The other was the White Tern (or Fairy Tern), a species found only in the Seychelles; it is pure white in colour, with a black bill and black eyes. These lovely slender birds roosted in the trees near the house; some of them incubated their single egg in the fork of a tree or balanced precariously on the upper surface of a branch.

Wood occupied his days swimming, fishing, gardening, collecting shells, hunting 'tec-tecs' (very small clams) in the sand at low tide, reading, and listening to music. He sailed over to Mahé to visit friends and to buy supplies – principally, tinned foods, bread and rice. Life on Cerf Island was simple and satisfying for Wood, although he missed having long, vigorous walks and perhaps felt rather lonely at times. Of particular importance in terms of his life on the island was the beginning of a long friendship with the Calais family, who in later years looked after him as if he was one of their family.[14]

But Wood was unable to settle, and suddenly he decided in mid 1932 to leave the Seychelles. Within a few months, he had sold his house and property on Cerf Island and departed for Africa. His friends, especially the Nageon and Calais families, thought they would never see him again. But in 1949, after an absence of nearly 17 years, he returned. He was by then 60 years of age – an age when most humans realise that life is not going to go on for ever and that the choice of a final home is necessary. At this time, he had started work as 'factor' at Sir Alfred Beit's house at Cape Maclear (see Chapter 8), and presumably he had enough money to contemplate building another home in the Seychelles where living expenses were comparatively low. Sir Alfred was evidently quite happy for Wood to divide his time between Cape Maclear and the Seychelles. So began the final years of Wood's life; he travelled backwards and forwards between Africa and the Seychelles, and Cerf Island definitely became 'home'.

In early 1949, Wood boarded the *Tairea* – the same ship on which he had travelled with Admiral Lynes 13 years previously – for the sea journey from Mombasa to Mahé. On arrival, he was met by Clement Nageon and his wife, with whom he stayed for a few days. He bought a small area of land on the northern side of Cerf, where he built another 'shack', very similar in design to the previous two that he had built in the Seychelles, and he rented an adjacent area from the Calais family (for only 60 rupees per year) so he could have a larger garden. The shack was near to the home of Samuel and Alice Calais and their sons, Marcel, Maxime and Daniel, and he was able to resume the close friendship with them that had begun in the early 1930s.

The shack was only a few metres above sea level. It was small and simple, built of wooden planks and boards, with a corrugated iron roof and concrete floors.[15] There was a central living room, a bedroom, a study and a simple kitchen. The doors were made of solid wood, and the large windows were wooden louvres without glass. The double doors from the living room led to a roofed veranda which looked out over the lawn to the beach fifty metres away, and to the island of St Anne and the northern tip of Mahé across the water. Furniture was minimal – just a few chairs, tables, desks, cabinets and a bed (with a mosquito net). At the back, and separate from the house, was the bathroom and two stone-and-concrete water tanks. One tank contained rainwater collected from the roof of the house, and the other contained stream water which was channelled along bamboo conduits from the nearby stream. Close to the sea, to the left of the house, was a boathouse. On the seaward side of the garden, the sand sloped steeply into the water; to the left and right, large takamaka trees overhung the sand, their spreading branches and dense wide leaves forming patches of deep shade.[16]

Wood always employed people to look after him and to do jobs for him; even with only a small income, this was quite possible because labour charges were so low. Throughout his lifetime, from his early days in Victorian England to his many years in colonial Nyasaland, this was the norm for people of his standing. On Cerf Island, he employed André Servina (whom he described as 'a very decent sort of old retainer') and his wife Emily. They lived in a little house which Wood built on the edge of the property, and looked after his everyday needs – cleaning, washing, and cooking simple meals. Wood spent a lot of time with the Calais family, especially with Marcel, who accompanied him on many trips to the reef to catch fish and hunt for shells, and who became a loyal and trusted friend, sharing many of his interests. Marcel also helped with his boats, shopping expeditions, and many other day-to-day activities. Wood became part of the Calais family, and on most days he joined them in their house for the main meal at midday, and very often they sent a simple supper to his 'shack' in the evening. Life on Cerf Island would have been much less enjoyable for Wood without the Calais family.

One of Wood's priorities when he had moved into his shack was to plant a new garden. He planted casuarinas on the boundary to provide a windbreak, and many ornamental trees commonly found in the Seychelles, such as calice de pape (*Tabebuia pallida*), takamaka (*Callophyllum inophyllum*), sandragon (*Pterocarpus indicus*), badamia (*Terminalia catappa*) and jamalac (*Syzygium samarangense*). In addition, there were coconut palms, breadfruits, mangoes, frangipanis, allamandas, and many other species. Some plants were grown from seeds brought from Cape Maclear. Wood evidently did not realise that all these trees would grow so quickly, and after about five years there was such a forest of vegetation that he was constantly pruning and lopping to allow sunlight to filter through to the orchids that he had placed on many of the tree trunks. The garden became a riot of colour and tropical exuberance, and was always in need of attention to prevent it becoming a jungle.

Life on Cerf Island (and in the Seychelles in general) was quiet and uneventful, in the nicest possible way. Each day was governed by the weather, the state of the tides, and personal inclination. There was never any hurry to do any particular thing on any particular day, and for most of the time decisions were made on the spur of the moment. If somebody unexpectedly turned up, fishing could wait until another day; if the sun shone and the tides were correct, then swimming and shell hunting took priority over a visit to Mahé. For Wood, life revolved around collecting shells, swimming, fishing, gardening, going to Mahé for 'shopperies' (as he called his shopping expeditions), visiting friends there, and enjoying the visits of numerous people who came to see him. One visitor was Alan Kohn, a biologist from America studying marine molluscs, who stayed with Wood in early 1958; he remembered that meals were very simple:

> Jan 19. The food at Cerf is interesting. We live off the land and the sea. Mostly fried cordonier (*Acanthurus*) with pumpkins, arrowroot, some string beans. Today, octopus for lunch. Café creole in the morning – its very strong. Diluted 1:4 with milk it is good. RCW tells me the local coffee is not grown here but is the dregs from Africa – beans that can not be otherwise sold. Roasted, or rather burned, here, the coffee produced is very good.[17]

Collecting shells was one of Wood's abiding passions on Cerf Island. He never tired of going out on the reefs hunting for shells, cleaning and cataloguing them, and placing them in the numerous cabinets that he had built for his collection. Most of the shells that Wood collected were cowries and cones; these are marine molluscs that produce some of the most beautifully coloured and ornate shells anywhere in the world. There are 190–200 species of cowries, of which about 65 occur around the Seychelles and in the western Indian Ocean.[18] The largest species have shells measuring about 100 mm length, and the smallest about 8 mm; all of them are rounded and slightly elongated, and the underside has a longitudinal opening (through which parts of the living animal extend to the outside of the shell), bordered on either side by delicate striations etched into the structure of the shell. Cowries have always been special to humans, and were used for trade and as 'money' throughout the tropical world before coin money came into general use. The colours and patterns of cowries are truly wonderful. In the living animal (and in a shell that has not been battered by sand and exposure after the death of the animal), the shell is shiny, almost as if it has been painted with lacquer. The colours range from pale salmon pink to various shades of cinnamon, golden brown and chestnut brown; many are marked with darker, richer blotches, spots and lines which form intricate patterns on the paler background. All of them are beautiful, and glow like gems. Cones are very different to cowries. As the name suggests, the shell is conical with a wide opening at one end (where the head and foot of the animal emerge from the shell). There are many sizes, shapes and patterns of cones. Some have ornate projections from the surface of the shell, others have smooth shells; some are colourful (in rather sombre browns, ochres and chestnuts) while others are almost white and faintly washed with colour. Cowries and cones live on the reefs, either within the coral itself or in the sand between patches of coral. Shell hunting was possible only at low tide when the water was shallow and the sea was calm. Wood, accompanied by his helpers – either Marcel, Daniel or Maxime Calais, together with any friends who happened to be visiting – spent many hours on the reefs hunting for cowries and cones.

There were several ways in which they hunted. To begin with, they had homemade wooden boxes with glass at one end and open at the other so that they could look down easily through the water. Later, they used goggles when these first came into general use in the 1950s. In sandy areas, they used huge wooden rakes to rake cowries out of the sand in much same way as raking soil removes stones hidden below the surface.[19] Another method was to chip away at coral heads to expose cowries hidden in the hardened structure of the coral. When they were far out on the reef, several kilometres from land, one of the collecting team always stayed in the boat, just in case any of the searchers got into trouble. Luckily this never happened, except that on one day a huge shark shadowed Daniel without him ever being aware of its presence; only when Marcel saw the shark and clapped his hands loudly did the shark swim away.

Wood was a very careful collector; he took only perfect specimens for his collection, and for trading with other shell enthusiasts. He preferred to take live molluscs, and then prepare the shells himself, rather than taking shells from animals that had died a natural death and were damaged or corroded by contact with sand. Once he had obtained a required number of good, unblemished shells of a particular species, he never took any more. He had a list of 'desirata', as he called them, and much of his time was spent looking only for these species.

The number of cowries in a particular section of reef, and the variety of species, varied throughout the year. In some seasons, many were to be found; in other seasons, hardly any at all. The larval stage of cowries (called veligers) are pelagic and drift many kilometres from where they were spawned. The larvae settle if they reach a suitable reef, and then develop into adult cowries which are sedentary for the remainder of their lives. Hence, even though the reef had been searched many times before, there was always a chance of finding new cowries. Wood showed great persistence in his hunt for shells; on any day when there was a low tide, sunshine and calm seas, he went on to the reef to hunt.

Preparation of the shells took a long time. After boiling (to kill the animal), the tissue inside the shell and around the top of the shell (which hides much of the shell in the living animal) had to be removed. Some of the cleaning was done by hand, but Wood much preferred to bury the shells in the sand, where many small sand-living creatures ate the tissues without damaging the surface and structure of the shell. The location of buried shells was marked with sticks; however on one occasion, a storm removed the sticks and Wood insisted on digging up most of the beach in front of his house until he eventually found the shells he had buried! Once cleaned, each shell was individually numbered and identified, and the place, date and details of the habitat were carefully recorded. He took immense care, and spent a lot of time cataloguing his collection because he knew that any specimen without data was of little value.

Another activity on the reef that Wood enjoyed was hunting fish with an underwater harpoon gun. Even in his 60s, Wood still retained his love of the hunt, and pitting his abilities against those of his prey. He never fished with nets or with a fishing rod (methods he probably considered unsporting). The reefs around Cerf Island are rich in fish, both in number of individuals and number of species. On many occasions, he missed most of the fish he shot at, but sometimes he succeeded. He always referred to the fish by their Seychellois name and their scientific name – names that sound wonderful when read aloud: kakatoi rouge (*Pseudoscarus dubius*), pampe (*Trachynotus ovatus*), raie chauve-souris (*Stoasodon narinari*), poule d'eau (*Platax pinnatus*), carange chasseur (*Carange speciosus*) . . . At other times, Wood just enjoyed being among the fish, watching them swimming in shoals, and marvelling at their beautiful colours.

Boats are an essential part of life in the islands. Over the years, Wood had a series of different boats. Although he did not sail on Lake Nyasa, he must have become a competent sailor of small boats in waters around Cerf Island. Each of his boats was given a name, each recalling something that was pleasurable to him. In 1931, his boat was *White Tern*, named after the beautiful elegant bird that has the freedom to fly effortlessly over the seas. In 1949, his small sailing dinghy was *Nyasa*, recalling the lake where he had spent so many happy years in Nyasaland. Later on, he purchased a larger 16-foot sailing boat with an outboard motor because he needed something bigger and more seaworthy for going backwards and forwards to Mahé; she was called *Aloha*. There was also a little canoe called *Ticoyo* which was used for paddling or poling around Cerf Island. One can only guess at the origin of this name, but most probably it is named after the young half-caste boy who is the hero of the book *Ti-coyo and His Shark*.[20] Ti-coyo is a delightful little boy living in the Caribbean who becomes friendly with a young shark he called 'Manidou'; but he was also a naughty youngster who frequently disobeyed his parents and did not like authority, especially when authority wanted to get rid of the shark. Ti-coyo had many exciting adventures in the sea with Manidou. But as the shark became bigger, he became more of a threat to humans swimming in the sea (but not to Ti-coyo, of course), and Ti-coyo encouraged the shark to catch swimmers who represented authority and those who profited from the misfortunes of others – especially developers and unscrupulous money-makers. But, besides eating swimmers, the shark also had lots of good characteristics, and on one occasion helped rescue a boatload of 'good people' from an island that was being destroyed by an erupting volcano. It is not difficult to see why Wood named his canoe after Ti-coyo – the little boy in the story had a lot of the same characteristics as Wood himself.

Visits to Mahé were a regular feature of Wood's life on Cerf. These were partly for business – posting letters, going to the bank and shopping – and partly for pleasure. He used to sail over early in the morning with Marcel in *Aloha*; if the winds were good, it usually took about an hour. They had breakfast at the Pirates Arms, which was Wood's favourite place to eat and to meet friends. The Pirates Arms was on the edge of the harbour and, from the windows, Wood could watch the boats on the sea between Mahé and St Anne.[21] After breakfast, Wood started his 'shopperies', visited the archives and the botanical gardens, and met friends. If he was staying for a few days at the Pirates Arms or with the Nageon family, Marcel returned to Cerf in *Aloha*, coming back for Wood when he was ready to return. On many occasions Wood had tea or lunch with Bax Bentley-Buckle, the manager of the Southern Line, and he frequently went to Beau Vallon, on the opposite side of Mahé, to visit friends who were also shell enthusiasts. At the archives, he studied old documents relating to the history of the Seychelles, and after visits to the botanical gardens he came away with new plants and orchids for his garden. Sometimes there were special events, such as the Queen's Birthday Garden Party, or a dinner at Government House. These visits were necessary, but nevertheless happy, interruptions to the peaceful life on Cerf Island.

During the wet season there were many days when it was impossible to go to the reef, and even gardening was impossible. On these days, Wood catalogued shells, wrote letters, kept his accounts up to date, read books and listened to his gramophone and radio. He must have been a prodigious letter writer; in these days there were no telephones, and contact with the outside world was slow. He must have planned his travels and arrangements very carefully, because wherever he ended up, friends were waiting to meet him and look after him. Most of his letters were handwritten in ink with a fairly broad nib; he had very distinctive, legible handwriting, easily recognisable on his specimen labels, notes and letters. Keeping accounts also occupied much time. His income was only about £300 per annum from investments, which, even in those days, was rather little to live on. Marcel Calais remembered how Wood was very careful with money:

> He didn't seem to spend money on things he did not want, or thought were useless. He used to think first. I remember one day when he wanted to buy a camera. He did not have enough money . . . he was very fond of sweets, but he cut out sweets to save enough to buy a camera. He was a man like that – he could deprive himself of something if he really wanted something else.[22]

Sundays were very special days. On every other Sunday, Georges and Lena Nageon and their family, and often their friends as well, sailed over to Cerf Island to visit Wood. They called him 'RC', but to the children he was always 'Mr Wood'. Lena Nageon retained happy memories of their Sunday visits. The whole day was spent swimming and diving, and then sitting in the shade of the takamaka trees for long conversations and eating lovely picnics which she had prepared. At times, they would go hunting for shells on the reef, wearing their goggles and flippers. Wood enjoyed these days immensely, especially the interesting and intellectual conversations – and there were few people in the Seychelles who had such wide interests and tastes as he had. Lena Nageon remembers that he could be quite critical about the Seychellois, and that he used to say that they suffered from 'coco-itis' – a name which he invented himself – because all they were interested in was coconuts. 'He was a very dear nice man . . . I used to make those little cakes that he liked very much – little tarts, you know, with jam. If we were there for the day, he often used to say "Now is the time to have grub!" He loved sweets, and always had a little store of them.'[23]

The Sunday visits to Cerf also made a lasting impression on the children of those days. Nadege Nageon remembers when Wood returned to the Seychelles in the late 1940s and the happy days she spent on Cerf Island:

As we became more familiar with it, Cerf Island grew on us, and we would often sail '*Pinta*' to a small sandy cove shaded by tall trees where we would spread our beach blanket in preparation for a veritable feast consisting of the usual picnic fare: saffron rice pilaf and corned-beef-and-onion pie, with trimmings . . . A few hundred yards off Cerf Island, in the most beautiful aquatic setting, RC was fated to re-enter our lives . . . Just as '*Pinta*' entered the bay leading to the cove, she was becalmed causing her to drift in the direction of a white lapstroke rowboat, behind which stood someone up to his neck in water, hatless under the sizzling tropical sun, and wearing goggles! We had no choice but to introduce ourselves to the stranger who we took for some eccentric or other from parts unknown. Foreign visitors were few in the Seychelles in those days and they were generally Englishmen who rapidly grew tired of the islands where amenities were lacking and foraging for food was one's daily concern . . . In a clipped English accent, he shouted his name across *Pinta*'s bow – Rodney C. Wood was once again in our midst!! An invitation was soon extended to him to join us for lunch at the cove, which he eagerly accepted.

RC (as we called him) was tall, lean, bronzed with bright blue eyes that sparkled when he heard a good joke. An aquiline nose and noble forehead gave him an air of distinction and authority. He walked some-what like a general 'on the eve of battle', posture perfect although he held his arms back somewhat when he walked, bent slightly at the elbow. He spoke in a commanding voice and had quite a presence. His hair was thin, light brown, and was beginning to recede slightly when we first knew him. His exuberance was contagious, and he enjoyed life on Cerf, especially his small cottage 'The Little Grey House' and, the symbol of his independence, 'Ticoyo' that took him to his shell-hunting areas in the bay. . . . RC always wore khaki shorts and was invariably bare-chested. When swimming he wore deep-blue swim trunks. When travelling, be it to Mahé or East Africa, he was always garbed in his 'safari outfit' for want of a better name – off-white shorts and short-sleeved jacket with large pockets, the whole held together by a belt of the same material buckled in front. During our first picnic with RC, it became evident that no future picnic would be complete or enjoyable without this friendly, charming and exuberant Englishman! It served as a prelude to our intro-duction to his bachelor pad within sight of the cove.[24]

On the next picnic day, the family was invited to Wood's cottage – a special treat for little Nadege:

We soon discovered that to be invited again to The Little Grey Cottage one had to prove oneself as a visitor by scrupulously removing sand from one's feet – beach sand that is sticky and messy. RC would lead the way, the index finger of the right hand pointing to a doormat placed at the front entrance. It was a modest cottage, with barely four small rooms, possibly fewer. In his own way, he was the perfect host . . . always gracious and likeable. [On one occasion] he allowed us to have 'only a peek' at his shell collection. He invited each one of us in turn to stand next to him . . . while he opened the drawers of a cabinet in his small living room to reveal his precious seashells lovingly wrapped in cotton wool. Each little shell was appropriately labelled with its name and number. We would have gladly lingered longer to have a good look at each and every one of them, but RC allowed only the briefest peek as though fearing that the shells would take fright and flight![25]

The Sunday picnics became a regular occasion. Each followed the same procedure and consisted of the same type of picnic. Nadege Nageon remembered:

> The corned-beef pie was made of corned beef from Argentina (via England) and was served in slices. The pilaf [eaten cold] was made from the best rice available at the time – Bengal rice in which slices of the local sausage figured, as well as black peppercorn. The pilaf was always accompanied by a chutney of fresh 'fruit de Cythère' (dubbed 'golden apples' by the English community). The 'fruit de Cythère' was chopped up while still green, and oil and vinegar added, as well as finely chopped onion. Often a hot chilli was added for an interesting flavour. No saffron-rice pilaf was complete without this chutney. . . . Dessert consisted of fresh fruit from the garden, such as paw-paw, mango and bananas. We drank fresh lemonade. How RC enjoyed these picnics in the shade! Every mouthful drew from him loud exclamations of surprise and satisfaction, and we never had to prod him towards the huge pot of saffron rice for a second helping. He was easy to please, and everyone strove to please Rodney! . . . During those Sunday outings, we never fished, and later in the day dug tek-teks (a very small clam) at the low-water mark. There were tek-teks on most beaches, and they made a delicious soup served in the tiny shells. They had to be allowed to expel the ingested sand (in salt water only) overnight, so as to have a good soup free of sand. We also spent a considerable amount of time swimming and diving. The water in the bay, as elsewhere around the islands, was crystal clear, always.[26]

Marcel Mathiot was a young boy in the 1950s who was given the nicknames of 'Marchello' and 'Tioté' by Wood. Marcel remembered that he used to help Lena Nageon with the preparations for lunch:

> . . . and of course she baked Rodney's favourite jam tarts. We used to get there very early in the morning, and would go out snorkelling, often fishing, and he would be raking for shells. And then after lunch, he would sit down and tell us stories about the old days, Nyasaland and so on. And of course, he always looked forward to teatime for his favourite jam tarts. At the time, 'Cerf Island was the Calais and Rodney'. Rodney

Wood was a bit of Cerf Island. He liked children and children liked him. I think he was a legend; he made a deep impression me.[27]

Another of these youngsters was Aubrey Michel, who remembers that on arriving at Cerf, he was always given a sweet by Wood, who would carefully collect the wrapper so it could not blow away in the wind. Like the others, he remembers Wood as:

> a tall man, brown as a berry, and with a tall 'military bearing'. He was pleasant, well-spoken and laughed easily. He had a large collection of shells in the house laid out in specially made chest of drawers. He was very fond and protective of animals, his house was always swarming with lizards. The only time I heard him raise his voice was when my mother's dog (a black dax) had a go at one of the lizards – he was quickly put back in the boat. One thing I do remember him trying to teach us was to float in the sea . . .[28]

Wood loved these picnics as much as his visitors did, and referred to the Nageons and their friends as the 'goggling gang'. He and the children had a lot of fun, pretending they were pirates on remote islands searching for treasure to be smuggled home at the end of the day; Cerf Island was the perfect place for such pretences, with its remote sandy beaches, little coves fringed with palms, blue sky and sea, and sailing boats moored in the shallows hidden from view. He often wrote about these occasions in his diary:

> 6.xi.55. Georges and Lena N came over after breakfast in Pinta . . . G and I swam and floated in the water singing songs, and then eating mangoes and pineapple, after which to my place for bananas and biscuits.[29]

And on New Year's Day 1956:

> What a grand day we had. After breakfast Georges and Lena arrived with a goggling gang: Armand, Duranto, John and Jack d'Umenville, their sister Seline, a delightful girl, and little Marcel Mathiot. They all goggled

in morning at the E corner of St Anne Is., where well sheltered from the SW to W wind which blew strongly all day, just outside the reef. After a wonderful picnic lunch on my veranda, we all went off goggling again to same place, returning for a late tea at 16.45. I got two fish and lost two more; also saw a small ray and a small shark. All very merry all day and much fun. A New Year's Day to remember with joy, and gratitude.[30]

For nearly twelve years, Wood travelled backwards and forwards between Africa and the Seychelles. His endless search for peace and quiet (which he found in the Seychelles) on one hand, and the call of Africa on the other, were as strong towards the end of his life as they were at the beginning. But at the end of his life, he regarded the Seychelles as 'home' more so than Africa, and always exclaimed when he returned to the Seychelles, 'Home again!' Although he lived in the Seychelles on eleven separate occasions, the total length of time during which he resided on the islands was only about nine years, compared with the thirty-seven years that he lived in Africa.[31]

Chapter 11

AFRICAN NATURALIST

Rodney Wood was a naturalist at heart and a born collector. His favourite occupations were being out of doors, hunting, collecting animals and then preparing and cataloguing them meticulously for his collections, and writing notes about their natural history. His love of natural history was evident when he was a young boy at Ardvreck School, and it continued throughout his life. Like many young boys, he was especially interested in butterflies and all the creatures that lived in rivers and ponds. It is very easy to imagine him on his country rambles with an insect net, and 'tiddling' in the streams with a jam jar on a string. Unlike most people, he did not lose this fascination with nature as he grew up. He also loved beauty, and hence his collections were mainly of the most attractive and colourful of animals – butterflies, birds, mammals and seashells. His earliest 'scientific' collection was sticklebacks which he collected at Harrow in 1905 when he was 16 (see Chapter 1), and the last was butterflies[1] collected when he was almost 72 and only a few months before his death. However, Wood's collections are only one aspect of his extraordinary abilities as a naturalist. He knew an enormous amount about all the plants and animals that lived around him. Brian Marsh, the crocodile hunter (see Chapter 8), for instance, wrote that Wood knew the scientific, English and Chinyanja names for every species,[2] and anybody who went for walks with him in the bush commented that he was a walking encyclopedia and would talk non-stop about the plants and animals that they encountered. Wood also realised that a universal appreciation of natural history and conservation was possible only if everyone had the opportunity to learn about nature: hence his numerous letters to newspapers, his informal talks to young and old, both African and European, and his desire to establish a natural history museum in central

Africa. His thoughts about museums developed very early on in his life, as he related in a letter to Oldfield Thomas during the First World War. After commenting that he had no private means and had to work on the cotton plantations for his living, he continued:

> [My] collecting is done in my spare time and only then if I can train local people to do most of the work. Natural History is however my greatest joy in life and I believe, and most people tell me so, my true 'calling'. It is my ambition to try to get somebody with sufficient cash or patriotic philanthropy to found a *national* Museum for Central Africa (either in N. Rhodesia or this country . . .) of which I might be appointed as curator at a small salary, and so be able to work it up and devote my life to C. African natural history. Collecting with me is almost a mania and it is only by the formation of a local museum that any really good work can be done out here [Wood's italics].[3]

Sadly for Wood, his dream of a museum with himself as curator never materialised. However, he did live to see the foundation of museums in Southern Rhodesia, Northern Rhodesia, and Nyasaland although for him they were established far too late.

During my research for this book, I wanted to trace all of Wood's collections. This had a twofold purpose: to document the animals that he collected in Africa, and to provide information on the dates and localities where he collected. The sources of information were specimens in museums (where the specimen labels provide the name of the species, the locality and the date of collection), published scientific papers (which referred to Wood's specimens either as individuals (such as the types for new species or subspecies) or as collections (such as his fish from Lake Nyasa or his small mammals from Chiromo), and his personal diaries and writings. These records provided a list of 516 localities, dating from January 1913 to August 1961. For each record, I listed the date (to the nearest date or month), the locality, the country, and the animal taxon. Wood's major collections are insects (of many sorts), fish, birds, mammals and seashells. He rarely collected other animals that are readily or easily seen, such as frogs, reptiles or dragonflies. There may have been good reasons for this, one of which concerned the method of preservation. Frogs and reptiles need to be preserved in a fluid preservative, a commodity that was extremely rare in central Africa for most of Wood's collecting years. All the animals that he did collect (with the exception of fish) could be preserved as dry specimens and kept in airtight boxes. Nevertheless, Wood still had considerable problems with the climate. In the hot dry season, specimens become brittle and easily broken, and in the wet season mould and fungi can quickly ruin any plant and animal specimen not kept in a dry airtight container. Sometimes he collected only where he was living, such as when he was supervising the cotton plantation at Chiromo, or working on his tea plantation at Cholo. At other times, he travelled ceaselessly in the pursuit of specimens for his collections. Wherever he was, he always took the opportunity to collect; during his travels with Admiral Lynes in South Africa and East Africa, he collected butterflies at many of the places that they visited for their studies on *Cisticola*, and on his way to and from the Seychelles he collected shells at each of the ports where the boats docked. The sign of a good collector is that he is always prepared; Wood was the perfect example of a good collector for he always took his insect nets and other equipment wherever he went. Tracing Wood's specimens has been a massive detective search, with many false leads, and although the majority of his specimens have been accounted for, some are still 'missing', while some others (especially insects) have not been individually recognised.

Wood's specimens are readily identifiable by the excellence of his labels. He wrote neatly and legibly in black ink, recording his field number and name, the locality and date, essential measurements, biological information, and the species name if he knew it. His writing is very characteristic and clear, and the information is complete and detailed. This is in contrast to many other specimens collected in Wood's day (and up to the present time) which are inadequately and poorly labelled. The carefulness of Wood's preparation and the excellence of his labels have ensured that his specimens are still of considerable value to modern researchers.

The search for Wood's specimens has been an endless source of enjoyment. It has involved lengthy correspondence with museums in England, America, Zimbabwe and South Africa. Best of all has been examination of his fish, birds and mammals in the Natural History Museum (the former British Museum of Natural History) in London, his butterflies in the National Museum of Zimbabwe at Bulawayo, and his shells in the Natal Museum at Pietermaritzburg.[4] His collections of mammals, birds, fish, insects and seashells were prodigious. This chapter describes Wood's collections and attempts to place them and his other accomplishments in a broader context, and assesses the influence that his work has had on central African biology.

Mammals

The first recorded mammal that Wood collected was a bat that he found in November 1913 at Chilanga in Northern Rhodesia, when he was working for the British Cotton Growing Association. The single specimen was a new species of slit-faced bat (genus *Nycteris*) and was named *Nycteris woodi*.[5] It was the first of many species that were named after him.

After his move to Nyasaland in 1914, Wood collected small mammals at Chiromo and Cholo. These years were very busy: from 1914 to 1918 he was supervising the cotton plantation at Chikonje near Chiromo, was a member of the Nyasaland Volunteer Reserve, and was involved with the war effort; and from 1918 to 1921 he was working on his tea estate at Magombwa near Cholo. He obtained his small mammals in several ways: he purchased some specimens from local people; he examined burrows and hollow trees; he caught bats which came into his house with his insect net; and occasionally he shot bats with his shotgun (normally used to collect birds). He did not have any mousetraps (until Thomas sent some from London). Helped by instructions and booklets sent by Thomas, he trained a local man to prepare his stuffed specimens while he made all the measurements and notes himself. The first specimens were not particularly good, but the later ones were excellent. Wood was anxious to have a reference collection himself, so he kept some of the specimens, while sending others to the Mammal Department of the British Museum of Natural History (BMNH) in London. The arrangement was that some of the London specimens would be returned to Wood after identification, and others should remain in the museum. Wood sent his specimens to London in separate consignments by parcel post (mostly during the war). There are 258 of Wood's small mammals in the BMNH, most of which are rodents and bats (see Chapter 4). However, in one of his letters to Thomas, in which he lists the number of each specimen in the consignment, the numbers go as high as 441, so it seems as if at least 200 specimens were kept by Wood himself.[6]

It is not known what happened to these specimens or whether they are still in existence. Several scientific papers were written on Wood's small mammals between 1914 and 1922; one describes the whole collection and others describe the three new species named after Wood.[7] Since then, the collection and these papers have been quoted in numerous scientific works on the mammals of Nyasaland/Malaŵi. Wood published a short paper listing the mammals of Chiromo and Cholo in 1949, with a caution about the reliability of local names for mammals.[8] Many of the Chinyanja names for mammals that Wood collected have been cited in other publications on mammals of Nyasaland.[9]

Wood hunted extensively in Africa. There is very little evidence about the larger mammals that he hunted. He probably started hunting big game when he first arrived in Southern Rhodesia in 1909, and by 1912 he owned his own copy of Stigand and Lyell's *Central African Game and its Spoor* (see Chapter 6). He acquired a very extensive knowledge of bushcraft and hunting, and became well known in local circles for his abilities. These indirectly led to his appointment at Game Warden, to his many discussions with the Chief Secretaries and Governors on matters concerning game, and to the invitation to write two chapters in Maydon's *Big Game Shooting in Africa*. His account of shooting Nyala is one of the classical stories of hunting in Africa, and it also indicates his abilities in bushcraft and ecology. Wood was very different to most hunters, in that he liked the skill of tracking as much as the shooting, he only killed when he was after a trophy or meat, and he liked to pit his abilities with those of his quarry on a 1:1 basis. On some occasions he shot with his bow and arrow, a feat that requires much more skill and knowledge of animals than shooting with a gun. Some of his specimens were sufficiently noteworthy that they were listed in Rowland Ward's *Records of Big Game* (see Chapter 6 of the present book). Wood continued hunting until he resigned as Game Warden and left Nyasaland in 1931; there are no records that he hunted thereafter.

Birds

Wood was an ornithologist for all his life. As with butterflies, his interest in birds probably began while he was at Ardvreck, where there were cases of stuffed birds in the school library, and bird nesting was undoubt-

231

edly a feature of the weekly 'Barvicks' (see Chapter 1). Wood's association with birds had three separate aspects. First, he collected many specimens of birds, as well as nests and eggs (especially during his early years in Africa); second, he also kept notes and records about the species he saw where he was living and during his travels; and third, he always liked to have birds close to his house and garden, so he attracted them by providing food tables and birdbaths.

During the period from 1917 to 1920, Wood collected 377 specimens of the local birds near Chiromo and Cholo, and a few from Domira Bay. Later on, in 1926 and 1928, he collected 39 specimens from near the Ludzi River, and 50 from the Lower Shire Valley. He kept records of each one of his specimens in a large foolscap-sized book (see Chapter 4). Altogether, his collection comprised 459 specimens.[10] Tracing the current locations of the bird specimens has been difficult because museums often exchange or sell specimens to other museums, and sometimes the records of these transactions are inadequate or non-existent. In 1917, Wood began corresponding with Dr W. L. Sclater at the Bird Room of the British Museum of Natural History, and after the war he sent specimens to London by parcel post. As for the mammals, he wanted to give some specimens to the BMNH, but wanted to keep others (after they had been identified) for his own reference collection. The BMNH has 99 specimens of Wood's birds in its collection,[11] and evidently the remainder of his collection remained at Magombwa while he was working for the Boy Scouts in 1922–1925. In 1929, he met Rudyerd Boulton who was collecting specimens of birds (and other animals) for the Carnegie Museum (see Chapter 6). Boulton persuaded Wood to sell most (if not all) of his specimens to the Carnegie Museum. It is difficult to know why Wood sold his collection to Boulton; at this time, he was the Game Warden of Nyasaland and had sold part of Magombwa to Arthur Westrop, so perhaps he was already having doubts about remaining as Game Warden, or perhaps he thought that by selling his collection he would have the financial freedom to return to the Seychelles. In the end, Wood was paid $125 for his birds, $70 for his nests and eggs, $340 for his insects, and $5 for his mammals, a total of US$560. When the 343 birds arrived at the Carnegie Museum in 1931, the Curator of Birds, Mr W. E. Clyde Todd, was displeased with many of Wood's specimens because, to him, the necks were somewhat elongated and did not come up to the high standards that he demanded.[12] Consequently, he decided to keep only 121 specimens (representing those species which were not in the museum's collection) and to exchange 220 specimens with the Cleveland Natural History Museum in Ohio.[13] Many years later, in 1954–1955, 29 of the Cleveland Museum specimens were sent to the Zoology Museum at the University of Michigan.[14] Thus, now most of Wood's specimens are housed in four major museums in Britain and the USA. It is true that some specimens are slightly elongated in the neck, but this is a minor fault, and most of them (when seen in a drawer alongside specimens prepared by other collectors) are not distinguishable by this fault; in fact, most of them stand out because they are so well prepared. At the time of their collection, there were very few specimens of birds from Nyasaland in any museum, and hence Wood's collection was (and still is) a valuable contribution to African ornithology.

Wood really enjoyed watching birds and recording their activities. His diaries are full of notes on the arrival of birds, what birds were nesting and other snippets of interesting information. Many of his observations were given to Charles Belcher, a keen amateur ornithologist, who incorporated them in his book *The Birds of Nyasaland*, published in 1930.[15] Birds are probably the best known of terrestrial vertebrates, and there have been many ornithologists in Nyasaland. One of the most prolific after Belcher was Con Benson; in addition to many scientific papers, he wrote *The Birds of Malawi*, which is still regarded as one of the definitive works on the subject. Benson utilised the observations of many ornithologists, including those of Wood. Wood himself wrote a number of short anecdotal articles on birds, many of which were published in *Ostrich*, the South African Journal of Ornithology (see Bibliography for list of Wood's publications); but he never wrote any scientific papers or published checklists of birds at the places where he lived. This is something that he could have done easily, since he had the knowledge as well as the ability to write good English. His copy of Belcher's book is richly annotated with locality records and comments, either adding information or disagreeing with something in the text!

Quite apart from collecting and observing them, Wood loved to have birds around him. At every place he lived, he had bird tables of food, special perches in the *khonde*, and birdbaths in the garden. The entries in his diary show the delight that he gained from seeing flocks of small birds bathing in the birdbath, swallows nesting under the eaves, and nesting birds in the shrubs. Many of the visitors to his 'shacks' commented that there were always birds feeding and drinking nearby. Just by his example, Wood encouraged everyone he knew to become interested in birds; and no doubt it was Wood's enthusiasm and knowledge that persuaded many of his friends to become amateur birdwatchers.

Fish

The first fish that Wood collected for which there is a record are the sticklebacks that he caught in the pond at Harrow in 1905 (see Chapter 1). In his early years in Africa, he collected four fish, of three species, when he visited Chiromo in 1913; at a later date he sent them to the British Museum. It seems as if these fish[16] were mixed up with Wood's 1920 collection from Lake Nyasa (described below), and were described by Regan (1921) as coming from Lake Nyasa. The mistake was not discovered until 1955, when Peter Jackson noted that these species were not known from the lake but were well known in the lower Shire. When Jackson asked Wood about these fish, Wood thought that the museum must have confused the 1913 collection with the 1920 collection and hence included the names of the three Chiromo species with the records for Lake Nyasa.[17] There is a cautionary tale to be learned from this story: Wood's Chiromo fish were sent without any labels (the current labels were written by someone in the museum), and so it is easy to understand how the two collections were mixed up.

At the request of Dr. C. T. Regan, Wood visited Lake Nyasa in 1920 to collect fish for the museum. Very little was known about the fish of the lake at this time; small collections had been made by Sir John Kirk during the Livingstone expedition, and by Sir Harry Johnston, Alexander Whyte, and Commander Rhoades (later captain of the *Guendolen*) – just enough to suggest that the lake contained a very interesting fish fauna. At the lake, Wood's collection was placed in a few big collecting boxes and inadequately labelled as 'Nyassa' (without any specific locality). However, in a letter to Dr Regan in August 1920, Wood wrote that all the fish had been collected at Domira Bay, 'a large shallow bay with a sandy bottom and with no rocks, mud or weeds'.[18] The 190 specimens were collected in 2–4 fathoms of water by drag nets and on baited lines.[19] When the collection (described by Regan as 'a very fine collection') was examined, it contained 46 new species, which more than doubled the number of species previously known from the lake;[20] with Wood's collection, the number of species was increased from 38 to 84, of which 79 were endemic. Regan named two of the new species after Wood: *Haplochromis woodi* and *Rhamphochromus woodi* (both large, widely distributed piscivore species which live in open water). Such a high level of endemism was most unusual and posed many questions as to how a lake could have so many species, and how such diversity could evolve. Regan recognised that Wood's collection indicated the presence in the lake of a rich and varied cichlid fauna, and arranged for Dr Cuthbert Christy to make more extensive collections on Lake Nyasa in 1925–1926 and on Lake Tanganyika in 1927–1928. Christy was a medical doctor who, in addition to many other activities, had been a member of the Uganda Sleeping Sickness Commission in 1902 and had participated in numerous medical expeditions in many parts of Africa.[21] He was also a keen naturalist and collector. Christy collected some 3,800 cichlids and 800 non-cichlid fish from all parts of Lake Nyasa.[22] It took many years for this collection to be examined and identified. As a result of the Christy collection, the number of species of fish known from the lake was increased to 220, of which 175 were cichlids and 45 were non-cichlids.[23] Many of the species of fish are an important feature of the diet of Nyasalanders, and much of the more recent research has been directed towards the nutritional value of the fish and the fisheries industry. The 1939 Nutrition Survey began studies on the important species of fish, but was interrupted by the war.[24] Studies were resumed after the war, first by Rosemary Lowe in 1945 to 1947, and later by the Joint Fisheries Research Organisation [JFRO] (see Chapter 8).

Research on the fish and fisheries of Lake Nyasa has continued, and has resulted in the publication of a very large number of scientific papers and management plans.[25] In recent years, studies on food webs, niche selection and evolution, phylogenetic relationships and behaviour incorporate ideas and methods that were unknown in Wood's day but that would clearly have fascinated him. The fish of Lake Nyasa (now thought to contain about 800 cichlid species, of which about 99% are endemic, and 45 non-cichlid species) provide one of the best examples in the world of how species evolve to produce such amazing biodiversity in a single ecosystem. Another aspect of the immense diversity of the lake is that many of the *mbuna* (small rock-loving species which feed on vegetation) are brightly coloured, have complex patterns of colours, spot and stripes, and are extremely beautiful; and hence many of them have become favoured aquarium fish throughout the world. They have been the subject of many lovely coloured photographs in books[26] and are well known in public and private aquaria. The two species named after Wood are not aquarium fish because they are too large and feed only on other fish.

Wood's main contribution to studies on the fish of Lake Nyasa is that he made the first major collection, which has been the catalyst for all subsequent studies. His collection showed that there was a great diversity of fish in the lake, and that most of the species were endemic. He did not make any further collections after 1920, although he participated informally in the work of Rosemary Lowe and the JFRO. The collections and studies of all these researchers and collectors have made Lake Nyasa (now Lake Malaŵi) justly famous, and the fish of the lake are cited as one of the most marvellous examples of evolution in freshwater aquatic ecosystems.

Butterflies

Of all the animals that Wood collected, butterflies were probably his favourite. He collected more butterflies than any other sort of animal, and he collected for the whole of his life. He loved the colour, intricate patterns and beauty of butterflies. Each butterfly that he kept was perfectly set and labelled (with scientific name, number, locality and date). Sadly, only one of his butterfly field notebooks has been found,[27] but he evidently kept written records of every butterfly in his collection. Wood had many special cabinets with drawers, as well as storage boxes, for his butterflies; most were made by African carpenters to his specifications, but some were purchased in England and sent to Africa by ship. The collection was made between 1915 and 1945, and is now housed in two museums. There are 3,600 specimens in the Carnegie Museum in Pittsburgh. These butterflies were purchased by Rudyerd Boulton in 1930 at the same time as he purchased many of Wood's other collections. The largest part of the collection, of about 7,000 specimens, is in the Natural History Museum of Zimbabwe in Bulawayo. As early as 1938, Wood was concerned about finding the most suitable museum to house his collection. During his travels through Southern Rhodesia in the 1930s, he visited the Natural History Museum and realised that (in the absence of a museum in Nyasaland) this museum would be the best home for his collection. But it was not until 1945 and 1946 that suitable arrangements could be made. Wood had hoped that the museum would pay him an annuity (rather than a lump sum) for his collection, thus providing an additional annual income. However, this never happened. At first, Wood's butterflies were housed as a separate collection in special cabinets (as he requested); but after his death, they were placed taxonomically within the general collection, as is the normal procedure in museums. In order that Wood's butterflies could be immediately recognisable, special blue labels printed with the words 'Rodney C. Wood Donation' were added to each specimen pin beneath the locality and identification labels.[28] When I examined Wood's butterflies in Bulawayo, it was extremely easy to spot them: holding a drawer of butterflies at an angle and looking sideways, the blue labels near the base of each pin were immediately obvious. Virtually every drawer of the huge research collection contained blue labels – a very clear indication of the size and comprehensiveness of Wood's collection. Another reminder of the importance of the collection, which I was delighted to see, was a large blue engraved notice on the inside of the main door which read:

THE RODNEY CARRINGTON WOOD COLLECTION
OF LEPIDOPTERA
[DONATED 1946]
IS INCORPORATED IN THIS COLLECTION
SPECIMENS ARE INDICATED BY BLUE LABELS

Wood's collection was primarily from Nyasaland, but it also included a substantial number of specimens collected in South Africa, Southern Rhodesia, Kenya and Tanzania during the 1930s when he was at Michaelhouse and when he was travelling with Admiral Lynes.

Wood's collection undoubtedly constitutes a major collection of butterflies from Nyasaland, partly because of its size and large number of species, but also because it is composed of specimens from many localities. George Arnold (the Curator of Invertebrates at the time, and later Director of the Museum) wrote that the butterflies were 'in perfect condition and well set'.[29] The collection has been used in many revisionary studies on butterflies of Africa, especially since it contains many types (i.e. the original specimen[s] from which new species and subspecies have been described). The collection in the Carnegie Museum (which Wood sold many years before the rest of the collection went to Bulawayo) is also of excellent quality:

> One can find material from Wood in virtually every major family when examining material from Africa. The specimens are very distinctive, having circular labels with characteristic large hand-lettering. . . . The material is uniformly in excellent condition, meticulously spread and well preserved. It tends to be synoptic, with one or two specimens of any given species. . . . There is no question that the entomological material deposited by Wood in the Carnegie Museum is of considerable diversity and scientific importance.[30]

Wood's collection and notes were used extensively for the standard work on the butterflies of Malaŵi by John Giffard; virtually every page mentions Wood's specimens and his notes on the food habits of caterpillars.[31] Wood often found caterpillars and kept them at home, feeding them on their natural food plant until they pupated and hatched and hence he was able to match caterpillar and adult.

One might have expected many butterflies to be named after him, but there appear to be only two: *Basiochila woodi* and *Charaxes xiphares woodi*. However, many of his specimens are cited as the type specimens, and given names that do not acknowledge Wood directly. Wood was not the only butterfly collector of his time; other names that come to mind (who were friends of Wood) were Ken Pennington in South Africa and John Handman and David Giffard in Nyasaland; collectively, and with others, they provided much of our knowledge of butterflies of central and southern Africa prior to the 1960s.

Other Insects

In his early years in Africa, Wood must have collected almost every small 'bug' that he could find. Although there are no notes or detailed records of his collection, the *Bulletin of Entomological Research* published lists of the insects that had been donated to the Commonweath Bureau of Entomology. Insects were received from keen collectors from all over the world, but especially from the countries within the British Empire. It is from these lists that the information on Wood's insects has been obtained. Between about 1914 and 1921, Wood sent 4,435 specimens to the bureau. Most were insects (4,351 specimens) but a few (84 specimens) were arachnids (spiders, ticks and mites). The majority were classified as 'Other Diptera' (which includes a vast array of flies), while others were listed more precisely, such as mosquitoes, dragonflies, beetles, fruit flies, bees, and 'biting flies'.[32] All these specimens are now integrated with the millions of specimens in the Entomology Department of the British Museum of Natural History. Since he was collecting in an unknown area of Africa, it is hardly surprising that Wood collected many specimens that later were described as new species. During this study, I have found references to 18 species that bear the specific name *woodi*,[33] but there may be others which have escaped my notice. It seems that after 1922, the Bureau did not publish lists of donations, and hence it is not possible to find out whether Wood donated further specimens after his return to Nyasaland in late 1925. During the 1920s, he worked closely with Dr W. A. Lamborn, the Chief Entomologist of Nyasaland, especially on the role of tsetse flies in the transmission of sleeping sickness and nagama.

Another collection of insects was purchased by the Carnegie Museum in 1930 (at the same time as the birds, mammals and butterflies referred to above). This collection comprised 900 diptera (flies) and 100 coleoptera (beetles).[34] Surprisingly, Wood did not show any particular interest in dragonflies; however, he did collect a few which were sent to the Bureau of Entomology in 1914–1915, and (at the request of Elliott Pinhey of Bulawayo) he collected several species at Cholo in February 1960. Pinhey referred to these specimens in his publication on the dragonflies of Malaŵi in 1966.[35]

Seashells

Wood's collection of seashells was very selective. He collected only cowries and cone shells (see Chapter 10), and ignored any other types of mollusc. In many ways, shells are very easy to collect because, once prepared (sometimes a rather messy and smelly business), they are easy to store and are not attacked by insects, pests or mildew. However, care must be taken to prevent them from becoming chipped or broken, but when placed in drawers (with cotton wool), they are relatively safe. Most of Wood's shells were collected in the Seychelles. The first collections were made in 1931–1932, although in those years more of his time was taken up with collecting fish, building his 'shack' and planting a garden;[36] the majority were collected from 1949 to 1961 (see Chapter 10). Shells were also collected from near Durban in 1933, on Mafia Island in October to December 1946, and at Mombasa, Beira and Mozambique during his travels between Africa and the Seychelles. The records of each shell (or of each group of the same species collected at the same time) were kept on card index files which recorded the date, place, size and condition of each shell.[37] The number of shells which Wood collected is uncertain, particularly since he traded shells with collectors all over the world. However, his major collection was given to Lady Courtauld, with the understanding that she would pass them on to the Umtali Museum when that museum was built. In the meantime, the shells were put on display at La Rochelle (see Chapter 9). After Lady Courtauld's death, the shells went to the Umtali Museum and later, when the National Museums of Rhodesia were reorganised, they were sent to the Natural History Museum in Bulawayo. Later, they were exchanged with the Natal Museum at Pietermaritzburg in South Africa, where other shell collections also existed, and where there was ongoing research in marine biology. The collection of some 1,000 shells is now part of the much larger collection of cowries and cone shells at Pietermaritzburg. Although Wood's shell collection was beautifully presented and labelled, it was rather biased in its scope and hence does not fully represent the mollusc fauna of the Seychelles.[38] Nevertheless, the collection is a valuable contribution to knowledge of the biology of marine life in the Seychelles, and is a (partial) record of what was there prior to 1960. One indirect effect of the collection is that the reefs around Cerf, St Anne, Round and Long Islands are well-known for their great diversity and are now protected within the St Anne Marine National Park.

Orchids and Other Plants

Wood was very fond of plants and gardening. He made many gardens during his life, and loved the colour and diversity of flowers. Tropical gardens are always full of flowers, attracting insects and birds in all seasons. Wood's gardening was rather haphazardous because he tended to plant trees much too close together and in time the garden became a jungle. Branches then had to be pruned, and trees removed. Wood took a great interest in having new and interesting plants in his gardens, and whenever he travelled, he brought cuttings and seeds home with him. He brought many 'new' plants into Nyasaland; one of the better known of his introductions was a large pure white frangipani with a yellow centre which he obtained from Mombasa (said to have come from Sir John Kirk's garden), and which is now widespread in Malawi. Over the years, he took plants from Nyasaland to the Seychelles, and from the Seychelles to Nyasaland. When he started to work as factor for Sir Alfred Beit, he became interested in orchids. Sir Alfred was an orchid enthusiast too, and together they amassed a wonderful collection in the garden at Cape Maclear. Wood kept detailed records of the orchids and of when they were in flower. He spent a lot of time looking after them, and he was always rearranging them so each could be in the optimum position for its own special requirements. Watering the orchids in the dry season was most time-consuming, especially when the pumps did not work. Cuttings from many of Sir Alfred's orchids were taken by Wood to the Seychelles, to Magombwa and to La Rochelle. There are still a few of Wood's orchids in the Seychelles – or at least the progeny of his orchids – and there is a magnificent collection of orchids at La Rochelle made by the Courtaulds (which include some of Wood's orchids and those of other collectors).

Conservation

Wood was a keen conservationist and had strong views about conserving animals, plants and ecosytems. This may sound a contradiction for someone who collected (and hence killed) so many animals for specimens. Undoubtedly Wood's views changed during the course of his life. Nevertheless, he viewed collecting as a means of understanding the diversity of life and the biological variation within and between species, and because of the importance of knowing what lived in a particular locality. Most of his collecting was for scientific reasons, and he never collected more specimens than was necessary. A collector who knows the fauna and flora can make sensible and correct decisions on what needs to be conserved, and which locations should be conserved. Many of the examples described in previous chapters show the influence that Wood had on conservation in Nyasaland. He had considerable influence on the decision to ban 'gun-boys', and on the legislation that prevented the wholesale slaughter of game animals to control tsetse flies (see Chapter 5). His period as Game Warden was important because he was instrumental in conserving selected areas that eventually became three of Malaŵi's major National Parks and Game Reserves (see Chapter 6). He realised that control of animals which were menacing local farms need not involve shooting, and that other methods such as vigilance, noise and protection were preferable. (There were, of course, some situations when shooting was still essential as in the case of man-eating lions and leopards and rogue elephants, for example).[39] Wood was also interested in the conservation of selected trees, and made the interesting suggestion that large historical trees should be conserved as national monuments.[40]

One of Wood's most long-lasting and important contributions (and perhaps the least known) concerns the introduction of 'reversed slope' and 'reversed bunding' into Nyasaland.[41] Wood saw this method of contouring slopes in order to prevent erosion and to increase the filtration of rainwater into the soil when he was in Ceylon and the Far East in 1925. He introduced 'reverse bunding' to his estate at Magombwa, and the method quickly spread throughout Nyasaland. It is a very noticeable feature of local farms in modern Malaŵi that maize, yams, cassava and other crops are planted across the slope, rather than up and down the slope, as seen in many other parts of Africa. This method of planting must have stopped the erosion of millions of tons of topsoil, thus stabilising the soil and retaining its richness and depth. Many of Wood's contributions to conservation in Nyasaland were not very obvious at the time but, with hindsight, his far-sightedness and acumen have been of immense importance to Malaŵi.

Natural History Museum

If Wood had had his own way, he would have been Curator of the Natural History Museum of Nyasaland (see Chapter 5). He floated the idea of a museum in Nyasaland in the 1920s, but there was neither the will nor the funds to build and maintain a museum at that time. Because Nyasaland was small, and not as wealthy as the Rhodesias, museums were established in Northern Rhodesia (Zambia) and Southern Rhodesia (Zimbabwe) rather than in Nyasaland. If Nyasaland had established a natural history museum in the 1920s or 1930s, it is certain that many of Wood's specimens (especially the butterflies and birds) would have stayed in Nyasaland rather than going to Southern Rhodesia and America. In about 1960, the Museums of Malaŵi, with a natural history section, were established; but by this date, Wood was over 70 and no longer a serious collector. He must have wished he was still a young man; how he would have loved to build up an impressive collection in the new museum!

Education

Wood was an excellent talker and raconteur, and he enjoyed showing his garden and birds to visitors. Walks in the bush with him were a fascinating experience because he was so knowledgeable. In this way, he imparted new information to many people, young and old, black and white. Wood's enthusiasm was infectious, and many youngsters became keen on natural history as a result of his enthusiasm. He often gave talks to groups about hunting, tracking and natural history. He spoke well, and must have been a good and inspiring teacher when involved with the Boy Scouts, when he taught at Michaelhouse and Livingstonia, and as Game Warden. Although some thought he was remote and stand-offish at times, he was extraordinarily good at interesting children in shells and butterflies (and in other aspects of natural history), and he often gave them a small souvenir (such as a beetle or butterfly) to take away.[42] He was equally at home with the Governors and Chief Secretaries of Nyasaland, and hence was able to give them his views on issues concerning animals and conservation. Although he did not write extensively, his letters and papers published in newspapers and journals were on topics of concern to him – in particular, the necessity to conserve, the importance of animals in ecosystems and an understanding of their benefits to humans, and an appreciation that all decisions concerning animals must be made on a scientific basis and not on emotion and hearsay.

Finally, how can one assess Wood's contributions to biology in Nyasaland and central Africa? The first point to make is that all Wood's collections are characterised by their excellence in presentation, and by the details of the information recorded with each specimen. Hence every specimen is of value, even though collected so long ago. Ironically, the value may increase with time because environments are changing so fast (mostly for the worst), and a historical record is the only evidence that a species did once occur at a certain place. Reay Smithers (formerly Director of the Museums of Rhodesia–Zimbabwe, and a great naturalist himself) knew Wood as a person and as a collector. He wrote: 'Rodney Wood was one of the kindest and most courteous men I have ever met and although he published very little during his life, his collections . . . and the meticulous data accompanying them have been of great value in taxonomic and ecological studies.'[43] Wood's collections have been useful in numerous ways: they have assisted with assessing biodiversity, biogeography and geographical variation of selected species and communities; they have been used to prepare guides and checklists; and they have stimulated further collecting and studies. Wood was interested in and knowledgeable about many groups of animals and plants, and was truly a 'naturalist' in the best sense of the word, whereas at the present time (when specialisation seems to be essential for a career in biology), most biologists have to specialise and become experts in a single field. The terms 'naturalist' and 'natural historian' were widely used in the past but have become debased in recent years, and are more likely to be used to describe a person who is an amateur rather than a professional. This is most unfortunate, because being a 'naturalist', in the way it is used to describe Wood, is a great compliment.

Wood lived at a time when it was possible to collect and observe before the great environmental changes of the twentieth century were evident. He would be horrified by the environmental degradation, loss of natural habitat, erosion, reduction in population numbers and over-exploitation of many species, especially mammals and fish, and at the enormous increase in the human population in Malaŵi which has taken place in recent years. He would have mourned the inability of humans (anywhere in the world) to balance the needs of human progress with the need to conserve the environment, on which all humans depend for a healthy and happy life. Many of his ideas where ahead of their time, and hence were not appreciated by his contemporaries. It was unfortunate that Wood did not express his ideas in a more scientific way and in places where they might have had greater influence and become known to a wider audience. But it was difficult in Africa in those days – there was no comparative material to help with identification, books and scientific publications were uncommon, and communication with the outside world was slow. Under such conditions, it is amazing how much Wood did accomplish. In other parts of eastern Africa, there were like-minded individuals who also collected, but sadly most of them did not record their experiences; one exception was Arthur Loveridge, who collected in German East Africa during the First World War and on later expeditions.[44] His adventures make fine reading, and describe the sort of life that he led and which must have been very similar to that of Wood.

Wood's collections will undoubtedly stand the test of time and will be consulted and quoted for many years to come. They will be referred to when taxonomic revisions are made, and they will be essential documentation for studies on geographical distribution and variation. The conservation of wonderful ecosystems in Malaŵi is due in part to his foresight. By any standards of comparison, he was a truly great African naturalist of the twentieth century.

EPILOGUE

After an absence of 11 months in 1960–1961, during which time he was in Africa, Wood arrived back in the Seychelles in September 1961 and resumed his normal life on Cerf Island. He wrote to the Westrops telling them that he had arrived safely, that the rains were good and that his orchids were flourishing.[1] A few weeks later, he wrote again telling them that he had been goggling for shells by himself in a rough sea, had got into difficulties, and had reopened an old rupture. The doctor in Victoria had advised that he should give up swimming and shell collecting. For Wood, there was no point in living on Cerf Island if he was unable to swim or collect shells; so he wrote to the Westrops asking if he could return to Magombwa and spend the rest of his days there among his beloved birds and butterflies. Arthur Westrop arranged passage on a boat in February 1962,[2] and Wood arranged for the letting of his house and garden on Cerf.[3] One day at about this time, when Marcel Calais was cutting his hair, Wood drew Marcel's attention to a lump on the side of his neck. They thought it was a mosquito bite and did not pay much attention to it. A few weeks later, at the very beginning of 1962, several more lumps appeared, and Marcel suggested that he ought to see a doctor in Mahé.[4] Wood refused to go to the hospital for observation, as the doctor requested, and spent the night at the Pirates Arms. But after a single night, his condition had deteriorated and he was admitted to hospital. He died three weeks later, on 29 January 1962, from lymph sarcoma,[5] just before he was due to return to Nyasaland.

So ended the life of one of Nyasaland's great characters and naturalists. Everyone who knew him remembers his generous nature, good manners and courteousness. They knew, as Lady Courtauld recalled, that he was at times an independent rebel and had a great dislike of officialdom.[6] He liked his solitude and was happy in his own company, but to those who knew him well he was a loyal and trusted friend. He loved fun and good music. He was an excellent shot and a good swimmer, and he had immense energy and enthusiasm. And as a naturalist, he was one of the best that has ever lived in Nyasaland and central Africa. He had a wonderful appreciation of the natural world, and enjoyed telling other people, especially children, about the wonders of nature. His influence spread far and wide, and his contributions to biology and conservation will long be remembered.

During this study, I have had to rely on records, natural history specimens, diaries and other people's recollections; sadly I never met Rodney Wood myself. Of all the reminiscences that I have received, one is particularly memorable as a final valediction: 'He fascinated and charmed young and old alike and was one of the few who became a legend in his own lifetime.'[7]

NOTES AND SOURCES

Chapter 1: Childhood in Britain: 1889–1909

1. Birth certificate for Rodney Carrington Wood. No. 73 of 1889, District of Paddington, Subdistrict of St John's, London (General Register Office, England); curriculum vitae, Rodney Carrington Wood, prepared for his application for Game Warden, Nyasaland (National Archives of Malaŵi, Zomba, Malaŵi).
2. Birth certificate for Alexander John Wood. No: 200 of 1860, Newcastle-upon-Tyne, Subdistrict of St Andrew (General Register Office, England).
3. Various sources, including: (a) Birth certificate for Edith Mary Carrington, Births in the District of Brisbane in the Colony of Queensland, no. 700 of 1861 (Birth: 31 August 1861. Parents: Charles Carrington and Jessie Carrington [née Bulgain]); (b) Queensland Consolidated Index of Births, National Library of Australia, Jessie Francis Carrington, born 26 April 1860; Edith Mary Carrington, born 31 August 1861; Ida Mary Carrington, born 6 November 1866; (c) New South Wales Pioneer Archives, National Library of Australia – Marriages: Reg. No. 1549, Chas Wilson to Jessie Carrington, 1879, Concord, Australia; (d) Recorded interviews and correspondence with descendents of Charles Carrington: Viv Fynn (Harare), Gwynneth Walker (England), Wyn Hooper (Harare), Anthony Hooper (Cape Town).
4. Marriage certificate, Alexander John Wood and Edith Mary Carrington, no. 79 of 1887, Christ's Church, Paddington, 15 February 1887 (General Register Office, England).
5. 1891 census.
6. See note 1, above.
7. 1891 census.
8. Marriage certificate, Alexander John Wood and Emily Blanche Wilde. No: 9 of 1895, Parish Church of St Saviour, Southwark. 2 February 1895. (General Register Office, England.)

9. See note 3, above.
10. Address used by Alexander Wood and Rodney Wood from about 1900 to 1936 (school records, Rodney Wood's cv, etc). Census data not yet available for these years.
11. Curriculum vitae, Rodney Carrington Wood (see note 1, above); London Trade Directories for the years 1901 to 1936.
12. Guildhall Library, Aldermanbury, London EC2 P2EJ. Contemporary photographs, telephone books and directories.
13. Burke 1940, Betjeman 1960.
14. Curriculum vitae, Rodney Carrington Wood (see note 1, above).
15. Jack Simmons Library and Archives, National Railway Museum, York; *Bradshaw's General Railway and Steam Navigation Guide*; contemporary notices, posters and photographs (courtesy: Philip Adams, Librarian).
16. Ibid.
17. Ibid.
18. Ibid.
19. Ardvreck School, Crieff; letters to author from Michael Kidd, deputy headmaster 1997-1998, and school prospectus. For history of Ardvreck School, see Ireland 1983.
20. See note 19, above.
21. Ardvreck School Magazine - various issues from 1899 to 1904; contemporary photographs. The songs sung at the concert in 1902 included 'De ole Banjo', 'Some folks', 'But it is so', 'Under the Willow', 'Who did', 'Darling Clo', 'Go Bye, bye', and 'Good-night'.
22. Ibid.
23. See note 19, above.
24. See note 21, above.
25. Ibid.
26. Ibid.
27. Harrow School; letters from School Archivist, 1995–1996; Harrow School Register [3rd Edition] 1800-1911 (ed. M.G. Dauglish & P.K. Stephenson); School Examination Records.
28. Ibid.
29. Anon. 1906.
30. See note 27, above.
31. Ibid.
32. Ibid.
33. Girling 1993; school photographs.
34. See note 21, above.
35. Anon 1933, 1940.
36. Letter in archives of the Ichthyology Department, British Museum of Natural History, London.

37. See note 1, above.
38. Apprentices 1891–1908, the Vintners Company (in Guildhall Library, MS 15214/4; see note 12, above).
39. Vintners Freedoms: June 1897 – September 1929 (in Guildhall Library, MS 15214/6; entry for 2 April 1914).
40. See note 1, above.

Chapter 2: Livingstone's Land

1. General books on the geography and landforms of Nyasaland/Malaŵi include: Debenham 1955; Pike & Rimington 1965; Ransford 1966; Tattersall 1982; Maurel 1990; Johnston & Garland 1993; and Carter 1987.
2. Foskett 1965; Coupland 1928.
3. Chadwick 1959. Other descriptions of the death of Bishop Mackenzie are given by Ransford 1966 and Martelli 1970. See also note 35 below.
4. Foskett 1965.
5. See note 3, above.
6. Ibid.
7. Foskett 1965.
8. Ibid.
9. Coupland 1928.
10. Foskett 1965; also see note 3, above.
11. See note 3, above.
12. Livingstone 1921; Ransford 1966.
13. Livingstone 1921.
14. Lt. E. D. Young, quoted in Reynolds 1997, pp. 8–9.
15. See note 3, above; also Jack 1990.
16. Jack 1900.
17. Hanna 1956.
18. Boeder 1980.
19. Perham 1956.
20. Moir 1923.
21. Ibid.
22. Hanna 1956.
23. Ibid.
24. Johnston 1897.
25. Johnston's map of Chiromo is now in the library of the Royal Geographical Society, London.
26. Maugham 1929.
27. Ibid.
28. Duff 1906

29.   Rees 1910.

30.   Colville 1911.

31.   Johnston 1897.

32.   Postcard. Department of Antiquities, Malaŵi.

33.   Baker 1971 (photo, p. 75).

34.   Mills 1911.

35.   Both Wilson 1936 and Mills 1911 write that the church of St Paul at Chiromo was built at Chiromo in 1906, and consecrated on 11 February 1907, close to the grave of Bishop Mackenzie. Chiromo is on the north side of the Ruo River, but the bishop was buried on the south side. At that time, the land around the Shire had not been designated to any foreign power, although after 1890, the grave was in the territory of Portuguese East Africa. Later (by 1907), a brick wall had been built around the grave. McDonald (1961) records that the grave was some 300 yards south of the Ruo, and east of the Shire River, and that the brick wall was still extant in 1961. An alternative story is that the remains of the bishop were re-interred in the new Church at Chiromo, and in 1922 they were moved again when the church was dismantled and taken to Blantyre (Stuart-Mogg 1994).

36.   Ransford 1966, p. 90.

37.   Jeal 1974, pp. 115–216.

38.   On her first trip, the *Pioneer* pulled the hull of Livingstone's second boat, the *Lady Nyasa*, and a barge. Other parts of the *Lady Nyasa* were placed on the *Pioneer* and on the barge. Livingstone's plan was to dismantle the hull below the Murchison Falls, carry all the parts 60 miles overland to the top of the falls, rebuild the boat, and then take it to the lake. The enterprise failed completely because it proved impossible to carry all the parts overland. In the end, the *Lady Nyasa* was rebuilt below the falls and never did any service on the lake or on the lower Shire. She steamed down the Shire with the *Pioneer* when the expedition left in 1864.

39.   Cole-King 1971, Reynolds 1997. Note that HMS *Pioneer* was a different vessel from the *Pioneer* that was used by the Livingstone expedition of 1861 to 1864 on the lower Shire and Zambesi rivers.

40.   Anderson-Morehead 1903.

41.   Ibid.

42.   Rees 1910 .

43.   Jeal 1974, pp. 115–216.

44.   Laws 1934, Livingstone 1921.

45.   Ibid.

46.   Ibid.

47.   Ballantyne & Shepherd 1968.

48.   Laws 1934, Livingstone 1921

49. Hanna 1956.
50. Hetherwick 1931. See also Buchanan (1885) for a good description of Blantyre and the Shire highlands in the days before the Protectorate was proclaimed.
51. Moir 1923
52. Cole-King 1973. Photographs of old Blantyre are printed in Lamport-Stokes 1989.
53. Ibid.
54. Hetherwick 1962.
55. Ibid.
56. Johnston 1923.
57. Johnston 1923; Maugham 1929. A history of 'The Residency' is given by Mell 1960.
58. Maugham 1929.
59. Ibid. Most the plateau of Zomba Mountain was planted with *Pinus patula* trees (as a commercial forestry crop) in the early days of the Protectorate. The grasslands are now restricted to parts of the 'rim' and the summits. The terms 'Zomba Mountain' and 'Zomba Plateau' are almost interchangeable. Zomba Mountain was the term mainly used in the older literature, whereas Zomba Plateau is preferred in modern maps and guides.
60. Maugham 1929.
61. Author notes.

Chapter 3: Early Naturalists in Nyasaland

1. Coupland 1928.
2. Foskett 1965.
3. Coupland 1928.
4. Foskett 1965.
5. Ibid.
6. Ibid.
7. Ibid.
8. Ibid.
9. Coupland 1928.
10. Accession book, Mammal Section, British Museum of Natural History, accession numbers 63.1.9.1 to 63.1.9.48. The numbers refer, in order, to the year (1863), month and day of accession, followed by the number given to the specimen.
11. Clendennen 1994.
12. Foskett 1965; Clendennen 1994.
13. Foskett 1965.

14. Ibid.
15. Coupland 1928; Chadwick 1959.
16. King 1993.
17. Ibid.
18. Chadwick 1959.
19. Johnston 1897.
20. Ansell & Dowsett 1988.
21. Huxley 1974.
22. King 1993.
23. Chadwick 1959.
24. Ibid.
25. King 1993.
26. Ibid.
27. Oliver 1964. These words are inscribed on a memorial plaque for Sir Harry in the Church of St Nicholas at Poling near Arundel in Sussex (author notes).
28. Oliver 1964.
29. Johnston 1897; Johnston 1923.
30. Baker 1971.
31. Johnston 1897.
32. Johnston 1923; Baker 1971.
33. Johnston 1897.
34. Ibid.
35. Oldfield Thomas wrote a series of five papers (all published in the *Proceedings of the Zoological Society of London* from 1892 to 1897) on the mammals of Nyasaland sent by Johnston and Whyte to the British Museum of Natural History. See Thomas 1892, 1893, 1894, 1896, 1897b.
36. Ballantyne & Shepherd 1968.
37. Thomas 1892.
38. Thomas 1896.
39. Thomas 1897b.
40. Citations for medals; Whyte: *Zoological Society of London* (1898); Johnston: *Zoological Society of London* (1895). See also Mitchell 1929.
41. See note 40, above.
42. Significant papers include Boulenger 1896, 1897, 1902, 1908; Butler 1895; Gunther 1864, 1893; Neave 1915; Sclater 1892, 1893, 1896; and Thomas 1897a, 1898, 1902. See also note 35, above.
43. Ansell & Dowsett 1988.
44. Ibid.
45. Johnston 1923; Baker 1971.

Chapter 4: First Years in Africa: 1909–1921

1. Rodney Wood, curriculum vitae for appointment to post of Game Warden (1928), National Archives of Malaŵi, Zomba.
2. McIntosh & Norton 1987.
3. Ibid.
4. See note 1, above.
5. Ibid.
6. Wood 1958a.
7. Bezzi, 1917, 1924.
8. Freeman 1957.
9. Parent 1935.
10. Andersen 1914. The type specimen of *Nycteris woodi* was collected at Chilanga, N.W. Rhodesia (now Zambia) in November 1913.
11. Wood 1958a.
12. See note 1, above.
13. Letters from Rodney Wood to Oldfield Thomas, 1916–1920 (now in archives of the British Museum of Natural History, London). Thomas's replies to Wood have not been traced.
14. Wood 1942; see also Happold & Happold 1998b.
15. Chikonje: John Trataris, personal communication, 1997.
16. Murray 1932.
17. Report of the Government Entomologist in Department of Agriculture Annual Report 1917 (National Archives of Malaŵi, Zomba).
18. Winspear 1960.
19. Rees 1910.
20. *Nyasaland Times*, various issues from 1913 and 1914.
21. Rees 1910
22. *Nyasaland Times* 1914.
23. See note 13, above.
24. Ibid.
25. Records in the *Bulletin of Entomological Research*, vols 4–12 (1914–1922). Total numbers of insects compiled by author.
26. Rodney Wood diaries.
27. Specimens and records in the Natural History Museum, Bulawayo, Zimbabwe.
28. Archives of the Wildlife Society of Malaŵi.
29. Accession book of the Mammal Section, British Museum of Natural History, London. Labels on Wood's specimens.
30. Kershaw 1922.
31. Ibid.
32. Ibid.
33. Wood's inscribed copy of Stigand & Lyell (1906), *Central African Game*

and its Spoor (now owned by Alan Foot, Harare); Roland Ward's *Records of Big Game* (referred to in a letter from Wood to Oldfield Thomas, 28 March 1915).

34. See note 13, above.
35. Letters from Rodney Wood to Dr Sclater, 1917–1920 (now in archives of the Bird Room, British Museum of Natural History).
36. See note 13, above.
37. Ibid.
38. See note 35, above.
39. Kershaw 1922.
40. Ibid.
41. See notes 13 and 29, above.
42. Ransford 1966, chapters 13 and 14.
43. Ibid.
44. *Nyasaland Times*, various issues in 1914–1915.
45. Bezzi 1918.
46. *Nyasaland Government Gazette*, 27 February 1915, Schedule no. 39 of 1915.
47. Murray 1932, pp. 428-429.
48. Charlton 1993.
49. Ransford 1966; Shepperson & Price 1958.
50. See note 1, above.
51. See note 13, above.
52. See note 1, above.
53. Details of the East African campaign are given in Moyse-Bartlett 1956 and von Lettow-Vorbeck 1922.
54. War Office 100/408; Public Records Office, Kew, England.
55. Murray 1932, pp. 428–429.
56. Ibid.
57. See note 13, above.
58. Ibid.
59. Hinton 1921.
60. On his printed stationery, Wood referred to his estate as 'Magombwa'. In Yao, the local language, this means 'bananas', and as Westrop n.d (p. 319) wrote, 'a very appropriate name for the little stream, on the banks of which grow so many varieties of that fruit.' While in Malaya, before settling in Nyasaland, Westrop printed his stationery with the name 'Magombe', which he assumed was the correct spelling. Apparently Wood (and Jim Harper, who looked after the estate during Westrop's absence) never forgave the change of name. The original spelling is retained here.
61. Rodney Wood diaries.
62. Westrop n.d. Author's notes on Magombwa 1994.
63. Archives Ichthyology Department, British Museum of Natural History.

Wood's first letter to Dr Regan was dated 16 November 1919, and there was a regular correspondence between Wood and Regan until mid 1920.

64. Boulenger 1909–1916.

65. Regan 1921; Accession book, Ichthyology Department, British Museum of Natural History, London.

66. Each box was 15 inches long, 12 inches wide and 14 inches tall. The sides were made of solid planks of wood, ¾ inch thick, joined by butt joints. Each corner was reinforced by metal right-angle flanges. On two of the sides were sturdy metal handles. The lid was secured by big brass hinges, and there was an inset lock. Inside the wooden container was a flush-fitting tin container, with a 6-inch circular opening on the top which had a screw lid and a leather washer. (Author's notes, Ichthyology Department, British Museum of Natural History, 1998.)

67. See note 13, above.

68. Ibid.

Chapter 5: Scouting and Planting: 1922-1929

1. Rodney Wood, curriculum vitae for application for appointment as Game Warden, 1928.

2. Boy Scouts (United Kingdom), Annual Report for 1922.

3. Boy Scouts of Canada, Annual Report for 1923.

4. Reynolds 1960, p. 107.

5. Ibid.

6. Ibid.

7. Don Potter, letter to author, 1995.

8. Richard Westrop (son of Arthur Westrop), personal communication, 1996.

9. Westrop n.d., pp. 18-19.

10. Boy Scouts of Canada, Annual Report 1924.

11. Rodney Wood, curriculum vitae for application for appointment as Game Warden, 1928.

12. Boy Scouts of Canada, Annual Report 1924.

13. Westrop n.d., p. 22.

14. Rodney Wood diary, 5 June to 7 December 1925.

15. Ibid.

16. Ibid.

17. Westrop n.d., pp. 25–34.

18. Ibid.

19. Hadlow 1939, 1960; Tea Association of Malaŵi 1986.

20. Murray 1932, p. 259.

21. Ibid.
22. Westrop n.d., pp. 25–34.
23. Ibid.
24. Rodney Wood field notebook for birds; accession lists in Carnegie Museum of Natural History, Pittsburgh, and Cleveland Museum of Natural History, Cleveland, Ohio.
25. Ibid.
26. Specimens in Mammal Section, Carnegie Museum.
27 Wood 1927a.
28. Wood 1929.
29. Wood 1926a.
30. Wood 1926b.
31. Cholo Planters Association. 14th Annual Report 1926.
32. See note 1, above.

Chapter 6: Hunter and Game Warden

1. Faulkner 1868, pp. 182, 184, 218, 221.
2. Drummond 1889, pp. 30.
3. Murray 1932, p. 268.
4. Johnston 1897.
5. Maugham 1929, pp. 97 et seq.
6. Duff 1906, pp. 140 et seq.
7. Dudley 1979.
8. Murray 1932.
9. Stigand & Lyell 1912. Wood's own copy, inscribed 'Rodney C. Wood, Chilanga, Northern Rhodesia' (see Chapter 4, note 33, above).
10. Dollman and Bulase 1928, Best and Raw 1973.
11. Wood 1932a (see also Maydon 1932).
12. Wood 1932b (see also Maydon 1932).
13. Wood 1932a (see also Maydon 1932).
14. Wood 1932b (see also Maydon 1932).
15 Johnston 1897, p. 377.
16. Ibid.
17 King 1993.
18. *Nyasaland Times*, 2 January 1913.
19 *Nyasaland Times*, 18 August 1913. The reference to askaris refers to a game control exercise when a detachment of askaris from the King's African Rifles was sent to clear game from an area with machine guns. They killed very few animals, and afterwards there was still plenty of game in the area.
20. Morris 1996.
21. *Nyasaland Times*, 2 January 1913.

22   *Nyasaland Times*, 2 March 1926; 12 March 1926.
23.   Ibid.
24.   Ibid.
25.   Gelfand 1964.
26.   de Guingand 1953, p. 48.
27.   Glasgow 1963.
28.   Ibid.
29.   Letters of appointment: Chief Secretary Nyasaland to R.C. Wood, 9 May 1929, 28 May 1929.
30.   Rodney Wood diary 2, 1 January 1929-31 October 1929; Rodney Wood diary 3, 1 November 1929-25 September 1933.
31.   See note 30, above.
32.   Ibid.
33.   Murray 1932, pp. 328–346. This is followed by the Game Regulations of 22 February 1927, pp. 347–368.
34.   Rodney Wood diary 2. Wood recorded the registration numbers of his vehicles and guns. For the vehicles, these were: Morris Cowley - CO-188; Douglas - LS 53. The vehicle registration codes indicate where the vehicle was registered: CO for Cholo and LS for Lilongwe. These low numbers are of interest because it shows that there were still not many vehicles in Nyasaland. His driving licence number was Cholo 160.
35.   Rodney Wood diary 2. Details of firearms: Rifles: Holland .375 (no. 28046, licence no. CO23); .375 Magnum (no. 388, licence No LL145); Lancaster .404 (no details). Shotgun: Greener 12 bore (no details). Revolvers: Wilkinson-Wembley .450 (service no. 4262, licence no. CO168); Colt .38 (no details); Colt .380 automatic (no. 6867) (Government Archives, Zomba).
36.   Rodney Wood letter. Government Archives, Zomba. Ref: 1/29.4 of 5 Dec 1930, plus minutes by Chief Secretary and Governor.
37.   Ibid.
38.   Dates from Carter 1987; see also Hayes 1967.
39.   Travel times as published in contemporary issues of the *Nyasaland Times*.
40.   See note 30, above.

Chapter 7: African Wanderings: 1932-1942

1.   Rodney Wood diaries: 1 November 1929–25 September 1933; 26 September 1933–11 November 1944.
2.   Ibid.
3.   Ibid.
4.   Letter by Wood in *Natal Mercury*, Durban, 17 June 1933.
5.   Editorial comment in *Natal Mercury*, Durban, 19 June 1933.

6. See note 1, above.
7. Dickson 1974. *Pennington's Butterflies of Southern Africa* was published posthumously (see Dickson and Kroon 1978). Some of Wood's butterflies are referred to in the text and illustrated in the plates.
8. See note 1, above.
9. *Michael's Chronicle* (journal of Michaelhouse School), extracts from 1934, 1935, 1936.
10. See notes 1 and 9, above.
11. See note 1, above.
12. Letter from Vaughan Winter to Ronald Brooks (Secretary of the Michaelhouse Old Boys' Club), 6 February 1995.
13. Letter from Donald Currie to R.E. Westrop, 12 July 1995; also letter to author.
14. See note 1, above.
15. Witherby 1943.
16. Ibid.
17. Diaries of Hubert Lynes (Bird Room, British Museum of Natural History, Tring, Herts).
18. See notes 1 and 17, above.
19. See note 17, above.
20. See note 1, above.
21. Ibid.
22. See note 17, above.
23. See note 1, above.
24. Cabinet: 'a polite and archaic term to mean closet (as in water closet or WC) or boudoir' (Oxford English Dictionary). See also note 17, above.
25. See note 17, above.
26. Ibid.
27. Smithers 1984.
28. See note 1, above.
29. Ibid.
30. Riley 1943.
31. See note 1, above.
32. Ibid.
33. See note 1, above.
34. Ibid.
35. See notes 1 and 17, above.
36. Witherby 1943.
37. Wood 1941.
38. Wood 1943.
39. Wood 1951. Archeological artefacts collected by Wood near Michaehouse and in northern Nyasaland are described and acknowledged by van Reit Lowe 1935, 1941.

40. See note 1, above.
41. *Nyasaland Times*; various issues, 1939.
42. Rees 1910 (see Introduction to 1986 reprint).
43. See note 41, above.
44. Ibid.
45. Ibid.
46. Ibid.
47. See note 1, above.

Chapter 8: Life by the Lake: 1942-1956

1. Rodney Wood diary 4.
2. Ibid.
3. Gunther 1864.
4. Jackson 1961, (see also Lewis *et al.* 1986, Konings 1990)
5. Cole-King 1971.
6. *Malaŵi News*, 27 July 1996.
7. Ibid.; Ransford 1966, chapter 15.
8. Ransford 1966, chapter 15.
9. Rodney Wood diary 4.
10. Bertram, Borley & Trewavas 1942. See also Lowe-McConnell 2006 for a very readable account of *Tilapia* in Africa.
11. Rosemary Lowe-McConnell, letters to author 1997–1998. Recorded interview with author, 20 August 1998.
12. See note 11, above.
13. Ibid.
14. Ibid.
15. Ibid.
16. Letter from Peter Jackson to author, 8 April 1995.
17. Rodney Wood diary 4.
18. Ibid.
19. Johnston 1897.
20. Maugham 1929, p. 57.
21. Potous 1956, pp. 70-71.
22. Medland & Van Zalinge n.d.
23. Wood 1937b.
24. Rodney Wood diary 4.
25. Ibid.
26. Ibid.
27. Records, Meteorological Service, Malaŵi
28. Rodney Wood diary 4.
29. Pike & Rimington 1965, Chapters 3 & 5.

30. Rodney Wood diary 4.
31. Pike & Rimington 1965, Chapters 3 & 5; Arnold 1952; Whitehouse 1952.
32. Arnold 1952; Whitehouse 1952.
33. Pike & Rimington 1965, Chapters 3 & 5; Arnold 1952; Whitehouse 1952.
34. Pike & Rimington 1965, Chapters 3 & 5.
35. Ibid.
36. Ibid.
37. Ibid
38. Ibid.
39. *Nyasaland Times*, 29 July 1946.
40. Ibid.; Cole-King 1982.
41. Rodney Wood diary 4.
42. Ibid.
43. Letter from Lady Beit to author, 1 February 1999.
44. Fort 1932.
45. *Who's Who*, various volumes.
46. Ibid.
47. Rodney Wood diary 4.
48. Ibid.
49. Letter from Peter Jackson to author, 8 April 1995.
50. Rodney Wood diary 4.
51. Ibid
52. Ibid.
53. Ibid.
54. Ibid.
55. Letters from Dr R. Wright, Medical Officer, to author, 8 March & 28 April 1996.
56. Letters from Don Baring-Gould to author, 2 November 1995 & 13 February 1996.
57. Ibid.
58. Letter from Christopher Barrow to Richard Westrop, 21 August 1995.
59. Letter from Brian Marsh to author, 13 May 1995.
60. Rodney Wood diary 4

Chapter 9: Magombwa and La Rochelle: 1958–1962

1. Rodney Wood diary 7.
2. Recorded interview with Richard Westrop, 15–16 September 1996.
3. See notes 1 and 2, above.
4. See note 1, above.
5. de Meillon 1930, Freeman & de Meillon 1953.
6. Raybould 1969a, b.

7. Lewis 1960; Lewis & Hanney 1965.
8. Hanney 1965. Additionally, Hanney wrote many other papers on mammals in Malaŵi.
9. Morris 1996, 1997.
10. Ibid.
11. Hayes 1972, 1978.
12. Documents sent by the Courtauld Institute, London.
13. *Who's Who* 1965, p. 674.
14. See note 12, above.
15. *The Search of Shangri-La*, a video about Sir Stephen and Lady Courtauld at La Rochelle, Penhalanga, Zimbabwe.
16. Notes made at La Rochelle by the author, September 1998.
17. See note 16, above.
18. Rodney Wood diary 7 and index cards.
19. Wood's shell collection, consisting primarily of shells from the Seychelles, but also of those from other parts of the world which he had acquired by exchange, was on display at the Umtali Museum for a few years. When the collections of the Museums of Rhodesia were being re-organised, Wood's collection was transferred to the Museum at Bulawayo, which was designated as the principal natural history museum in the country. Later, when it was decided that marine shells were not a priority in the museum, they were exchanged/sold in an arrangement with the Natal Museum at Pietermaritzberg in South Africa, where they have been integrated with the other collections of marine shells.
20. Letter from Mrs Gwynneth Scrivener (formerly Mrs Ireland) to author, 1995.
21. Ibid.
22. Newspaper cuttings and other documents in the Courtauld File, Mutare Museum, Zimbabwe.
23. Ibid.
24. Ransford 1966, chapter 16, plus other sources.
25. Ibid.
26. *Nyasaland Handbook* 1932, p. 76.
27. Pike and Rimington p. 133. Population numbers of humans in Nyasaland/Malaŵi have continued to increase since 1960, as have those of all other African countries. In 1990, the census recorded a population of about 8.2 million, nine times that in 1912 when Wood first arrived in Nyasaland. In 2008, the population was estimated to be about 13 million.
28. Westrop 1962.
29 Ibid.

Chapter 10: The Seychelles: 1921–1962

1. Much of this chapter is based on Wood's diaries, the author's notes made during a visit to the Seychelles in 1998, interviews with Marcel Calais, Lena Nageon, Marcel Mathiot and others in the Seychelles in 1998, and correspondence with Nadege Sylvia (née Nageon), Aubrey Michel, Armand Nageon and others. Since many interviewees and correspondents made similar comments about Rodney Wood and life in the Seychelles, individual credits were not possible for much of this chapter.

2. Rodney Wood diaries. There are few details of Wood's activities from July 1960 onwards, and Diary 10 has not been traced. Dates and places of residence, and dates of travel were taken from the index cards he kept.

3. Lionnet 1972, chapters 4, 5, and 6; Skerrett & Skerrett 1998. There are many other good guidebooks to the Seychelles.

4. Ibid.

5. Ibid.

6. Ibid.

7. Ibid.

8. Rodney Wood diaries; Lionnet 1972, chapter 1; Vine 1989.

9. Ibid.

10. Rodney Wood diaries.

11. Ibid. Details of Wood's singing voice has not been recorded.

12. Letter from Nadege Nageon to author, 1 September 2000.

13. Rodney Wood diaries.

14. Marcel Calais, personal communication. See also note 1, above.

15. Notes made by author in the Seychelles, September 1998.

16. Ibid.

17. Diary kept by Dr Alan Kohn, University of Washington, USA.

18. Lorenz & Hubert 1993. Numbers of species of cowries derived from maps in this publication.

19. Rodney Wood diaries.

20. Richer 1951, 1954.

21. The present Pirates Arms is on the same site as the original. However, it is no longer on the edge of the harbour. It is now several hundred metres inshore because the sea has been reclaimed to provide more building space in Mahé and for a new harbour. The modern Pirates Arms overlooks other modern buildings and no longer has the ambience of the Pirates Arms that Wood knew.

22. Recorded interview with Marcel Calais, Cerf Island, Seychelles, September 1998.

23. Recorded interview with Lena Nageon, Victoria, Seychelles, September 1998.

24. Letter from Nadege Nageon to author, 1 September 2000.
25. Ibid.
26. Ibid.
27. Recorded interview with Marcel Mathiot, Victoria, Seychelles, September 1998.
28. Aubrey Michel, letter to author, January 1999.
29. Rodney Wood diaries.
30. Ibid.
31. Calculations by the author, based on Wood's itineraries between 1911 and 1962 (Seychelles, 8 years and 8 months; Canada and UK, c.4 years; Africa c.39 years).

Chapter 11: African Naturalist

1. Hancock (1984) – butterflies collected 4 August 1961. The only period since 1913 when Wood did not collect was from 1922 to 1925, when he was in Britain and Canada with the Boy Scouts.
2. Letter from Brian Marsh to author, 13 May 1995.
3. Wood letter to Thomas, 9 April 1917. Archives, Natural History Museum, London.
4. Author's notes and records of Wood's collections. Wood's birds and butterflies in museums in the USA have not been examined personally.
5. Andersen 1914 – type description of *Nycteris woodi*.
6. Letters from Wood to Thomas, *Archives of the British Museum of Natural History*, vols. 20, 21 and 22. There is no record for the unaccounted specimens; likewise, any field notebook for small mammals that Wood may have kept has not been traced. It may be that some of the numbers were allocated to large mammals shot for other purposes (e.g. for trophies or food). Fifteen specimens collected by Wood are in the Carnegie Museum of Natural History in Pittsburgh, USA. These include 11 specimens from Chiromo and Cholo collected in 1918 and between 1926 and 1928, two collected at the Bua River in 1928, and two without locality records (Duane Schlitter, personal communication, 1994).
7. Papers describing Wood's collection of small mammals from Chiromo and Cholo are: Kershaw 1922 (complete collection); Hinton 1921 (type of *Uranomys woodi*); and Thomas 1917 (type description of *Scotoecus woodi*). Other publications on small mammals which refer to Wood's collections include: Ansell & Dowsett 1988; Hanney 1965; Happold & Happold 1997, 1998a,b).
8. Wood 1949d.
9. Chinyanja names for mammals have been published in Mitchell 1946, Sweeney 1959, and Hayes 1978.

10. At present, the 459 specimens are located as follows: 99 in the Bird Room, British Museum of Natural History (see note 11, below); 343 in American museums (see note 12, below); 17 were either given to 'W.P.Y.' (probably Rev William P. Young of Livingstonia) or were damaged in some way and were not kept (Wood's bird collection book). Benson & Benson (1977, p. 221) recorded information on the specimens in the Carnegie and Cleveland museums, but did not have such detailed notes as have been used in this book. Complete lists of Wood's birds (compiled from museum records and Wood's collection book) have been deposited in the museums listed here. Now, nearly 100 years since Wood arrived in Nyasaland, his work on birds is still remembered and quoted (Dowsett-Lemaire & Dowsett 2006).

11. Bird Room, BMNH. There are 99 specimens of Wood's birds, numbered 1920.10.25.1-14 and 1920.12.24.1-85. All were donated to the museum in 1920. All specimens have remained in the BMNH; none have been sold or exchanged with other museums.

12. Records and documents from the Wood file at the Section of Birds, Carnegie Museum of Natural History, Pittsburgh (courtesy of Dr Kenneth Parkes, Curator Emeritus of Birds). Of the 343 specimens that Wood sold to the Museum, the present location of the specimens is as follows: Carnegie Natural History Museum – 121; Bell Museum (University of Minnesota) – 1; Los Angeles County Museum – 1; Cleveland Natural History Museum – 191; and Zoology Museum, University of Michigan – 29.

13. Records and documents from Cleveland Museum of Natural History (courtesy of Dr Timothy Matson, Curator of Vertebrate Zoology). See note 12, above.

14. Records from the Museum of Zoology, University of Michigan (courtesy of Dr Janet Hinshaw). The 29 specimens were originally in the Cleveland Natural History Museum. See note 12, above.

15. Belcher 1930. Belcher was Attorney-General (and later a Judge of the High Court) of Nyasaland from 1920 to 1927. He then went to Cyprus (1927-1930) and to Trinidad and Tobago (1930–1936), and retired to Kenya in 1937.

16. The fish were labelled 1921.9.6.50 (*Eutrophius depressirostris*), 1921.9.6.51 (*Malapterurus electricus*) and 1921.9.6.65-66 (*Anabas multispinis*).

17. Jackson 1961, pp. 537, 543. There is still some confusion about exactly where and when Wood caught these specimens.

18. Letter from Wood to Regan, August 1920 (archives of the Ichthyology Department, British Museum of Natural History). The accession numbers of Wood's fish in the Ichthyology Department are 1921.9.6.1-226. Most of these are the 1920 collection from Domira Bay, but also included are the four specimens from Chiromo (see note 16, above), and thirty-

two specimens (of three species) from Nswadzi River, Cholo. Wood also presented a collection of 35 specimens (comprising 27 species) of fish from Cerf Island, Seychelles (accession numbers 1932.7.29.1-35). However, in his diary Wood listed 69 specimens identified to species level (plus 6 species identified only to genus level) which he collected in the Seychelles in 1931–1932, with a note that they were identified by J.R. Norman. The differences in the number is unexplained.

19.　Ibid.

20.　Regan 1921.

21.　*Who Was Who* 1929-1940.

22.　Localities where Christy collected (as given on labels on his specimens) include Deep Bay, Karonga, Vua, Fort Johnston to Fort Maguire, Monkey Bay, 'South end L. Nyassa', N. Kudzi Bay, and 'Bar to Kudzi'. Details of the collection were published by Trewavas (1935) and Worthington (1933).

23.　Bertram et al. 1942 provides lists of all the cichlid species.

24.　See note 17, above.

25.　There are literally hundreds of papers on the fish of Lake Malaŵi, far too many to list here. See Lowe-McConnell 2003 for a recent review (and papers listed therein).

26.　A good example is the magnificent book, with many colour photographs, by Konings (1990).

27.　This field book is in the Department of Entomology, Natural History Museum, Bulawayo, Zimbabwe, and records butterfly numbers 6001-7360 (24 August 1933 – end of 1944).

28.　Notes on the Wood collection of butterflies from the Department of Entomology, Natural History Museum, Bulawayo, Zimbabwe (courtesy of Karen Donnan, Honorary Curator of Entomology).

29.　Ibid.

30.　Letter to author from Dr John Rawlins, Curator of Invertebrate Zoology, Carnegie Museum of Natural History, Pittsburgh, 10 April 1995. Wood's collection of insects in the Carnegie Museum comprises about 3,600 butterflies, 900 Dipteran flies, and about 100 beetles.

31.　Giffard 1965, and notes from Dr John Wilson (Zomba, Malaŵi). Riley (1943) described *Baliochila woodi*, and Van Someren (1964) described *Charaxes xiphares woodi* and recorded additional specimens in 1972.

32.　Wood's collections in Northern Rhodesia and Nyasaland (except for butterflies) in 1913–1922 comprised 4,435 specimens. The principal groups of insects were: flies (2,940 specimens); beetles (724); bees and wasps (287); mosquitoes (172); and fruit flies (132). Figures calculated by author from records in the *Bulletin of Entomological Research*.

33.　For species of insects named after Wood, see Appendix 1.

34.　See note 30, above.

35. Pinhey 1966.
36. Rodney Wood diary 3.
37. Letters to author from Dr R.N. Kilburn, Natal Museum, Pietermaritzburg, South Africa, 1996, 1998, 2001; also documents from Mutare Museum, Zimbabwe (see Chapter 9, notes 16 and 19). In his will (dated 1 November 1961), Wood bequeathed a smaller collection of shells, which he had in his home on Cerf Island, to the Natural History Museum in the Seychelles (whenever such a museum was built). Although a museum was built a number of years later, the collection of shells remained, at the time of writing, in the possession of a private collector in the Seychelles.
38. Ibid.
39. In the early days, lions were extremely numerous throughout Malaŵi, and in some years many local people were killed by them. Interesting accounts of lions and other large game in Nyasaland and Malaŵi are given by Carr (1965), Muldoon (1955) and Morris (1995).
40. Wood 1949b.
41. Westrop 1962. See also Chapter 5.
42. Lady Courtauld; part of an obituary for Rodney Wood in *La Rochelle Bulletin*, 1962.
43. Smithers 1984; see also Pinhey 1962.
44. Arthur Loveridge wrote three books on his experiences and on collecting in Africa, all full of interesting reminiscences: *Many Years I've Squandered* (1949), *Tomorrow's a Holiday* (1951), and *I Drank the Zambesi* (1953).

Epilogue

1. Westrop n.d., p. 455.
2. Ibid.
3. Registration Division, Department of Legal Affairs, Republic of Seychelles: entry of 27 December 1961.
4. Interview with Marcel Calais, September 1998.
5. Death certificate, Seychelles. Civic status register no: 36-1962C. Cause of death was lymph sarcoma (cancer of the lymph glands). Wood did not die from injuries caused by trying to save someone in the sea, as Arthur Westrop wrote in his obituary in the *Nyasaland Journal*. The journal's policy was not to publish obituaries, but a special exception was made for Rodney Wood.
6. Lady Courtauld, *La Rochelle Bulletin*.
7. Ibid.

APPENDIX 1

Species named after Rodney Carrington Wood

Mammals

Uranomys woodi [now *Uranomys ruddi woodi*] (Rodentia, Mammalia)
Hinton, M. A. C. 1921. Some new African mammals. *Annals and Magazine of Natural History* 9(7):368–373. [Specimen, an old female, collected Cholo, Nyasaland, 2700 ft, on 27 October 1917. BMNH Accession number 21.2.16.1. RCW field collection number: 280.]

Scotoecus woodi [now *Scotoecus albofuscus woodi*] (Chiroptera, Mammalia)
Thomas, O. 1917. A new species of genus *Scotoecus*. *Annals and Magazine of Natural History* 8(19): 280–281. [Specimen, adult male, collected Chiromo, Nyasaland, 200 ft, on 2 October 1916. BMNH Accession number 17.2.1.1. RCW field collection number: 173.]

Nycteris woodi (Chiroptera, Mammalia)
Andersen, K. 1914. A new *Nycteris* from N. E. Rhodesia. *Annals and Magazine of Natural History* 8(13): 563. [Specimen, adult, collected Chilanga, N. W. Rhodesia, 4100ft, November 1913. BMNH Accession number 14.4.22.2.]

Fish

Haplochromis woodi [now *Stigmatochromis woodi*] (Cichlidae, Pisces)
Regan, C. T. 1921. The cichlid fishes of Lake Nyasa. *Proceedings of the ZoologicalSociety of London* 21: 675–727. [Six specimens collected 'Nyassa'. Accession numbers 1921.9.6. 139-144. Date of collection and locality not recorded, but probably Domira Bay.]

Rhamphochromis woodi (Cichlidae, Pisces)
Regan, C. T. 1921. The cichlid fishes of Lake Nyasa. *Proceedings of the Zoological*

Society of London 21: 675-727. [Three specimens, collected 'Nyassa' [BMNH Accession numbers 1921.9.6. 214–216. Date of collection and locality not recorded, but probably Domira Bay.]

Butterflies

Colotis ione f.woodi (Pieridae, Lepidoptera)
Original description not known.

Teriomima woodi [now *Baliochila woodi*] (Lycaenidae, Lepidoptera)
Riley, N. D. 1943. A new African Lycaenid butterfly. *The Entomologist* 76: 225–226. [Specimens, 3 male and 3 females collected Mt Mlanje, Nyasaland, 2500ft, 12 February 1938.]

Charaxes xiphares woodi (Nymphalidae, Lepidoptera)
van Someren, V. G. L. 1964. Revisional Notes on the African *Charaxes* (Lepidoptera: Nymphalidae) Part II. *Bulletin of the British Museum of Natural History (Entomology)* 15(7): 181–235. [Specimen, 1 female, collected Cholo, Nyasaland, May 1928.]

Other Insects

Aedes (Stegomyia) woodi (Culicidae, Diptera)
Edwards, F. W. 1922. Mosquito Notes. - III. *Bulletin of Entomological Research* 13: 75–101. Specimen, 1 female, collected Cholo, Nyasaland, 10 May 1916.]

Simulium woodi (Simulidae, Diptera)
de Meillon, B. 1930. On the Ethiopian Simulidae. *Bulletin of Entomological Research* 21: 185–200. [Specimens, 4 females, collected Cholo, Nyasaland, 14 September 1917.]

Chelyophora woodi Bezzi (Brachycera, Tephritidae)
Bezzi, M. 1924. Further notes on the Ethiopian fruit-flies, with keys to all the known genera and species. *Bulletin of Entomological Research* 15: 73–118. [Specimens, 1 male, 1 female, plus others collected Cholo, Nyasaland. October 1919.]

Bactropota woodi Bezzi (Brachycera, Tephritidae)
Bezzi, M. 1924. Further notes on the Ethiopian fruit-flies with keys to all the known genera and species (Continued). *Bulletin of Entomological Research* 15: 121–155. [Specimens, 1 male, 1 female collected Ruo, Nyasaland, 200ft. 15 September 1916.]

Carpophthoromyia woodi Bezzi (Brachycera, Tephritidae)
Bezzi, M. 1924. Further notes on the Ethiopian fruit-flies, with keys to all the

known genera and species. *Bulletin of Entomological Research* 15: 73–155. [Numerous specimens collected Cholo, Nyasaland, 2600ft. 4 October 1919.]

Eutretosoma woodi Bezzi (Brachycera, Tephritidae)
Bezzi, M. 1924. Further notes on the Ethiopian fruit-flies with keys to all the known genera and species (Continued). *Bulletin of Entomological Research* 15: 121–155. [Specimens, 1 male, 1 female, collected Cholo, Nyasaland. 16 November 1919.]

Ocnerioxa woodi Bezzi (Brachycera, Tephritidae)
Bezzi, M. 1918. Notes on the Ethiopian fruit-flies of the family Trypaneidae other than *Dacus* (s.l.) (Dipt.) -II. *Bulletin of Entomological Research* 9: 13–-46. [Numerous specimens collected Limbe and Chiromo, Nyasaland. 22 September 1916]

Pliomelalena woodi Bezzi (Brachycera, Tephritidae)
Bezzi, M. 1924. Further notes on the Ethiopian fruit-flies with keys to all the known general and species (Continued). *Bulletin of Entomological Research* 15: 121–55. [Specimen 1 male, collected Nyasaland. 4 October 1919]

Dacus woodi Bezzi (Brachycera, Tephritidae)
Bezzi, M. 1917. New Ethiopian fruit-flies of the genus *Dacus. Bulletin of Entomological Research* 8: 63–71. [Numerous specimens collected Chiromo, Nyasaland. 22 September 1916.]

Trypanea woodi Bezzi (Brachycera, Tephritidae)
Bezzi, M. 1924. Trypanéides d'Afrique (Dipt.) de la collection de Muséum National de Paris (suite) [Concl.) *Bulletin de Muséum National d'Histoire Naturelle, Paris* 1924: 88–91. [Numerous specimens collected Cholo, Nyasaland. No date.]]

Dryxo woodi Cresson (Cyclorrhapha, Ephydridae)
Cresson, E. T. 1936. Descriptions and notes on genera and species of the dipterous family Ephydridae. II. *Transactions of the American Entomological Society* 62: 257–270. [Specimens, 1 male, 3 females collected Cholo, Nyasaland. No date.]

Chironomus woodi Freeman (Nematocera, Chironomidae)
Freeman, P. 1957b. A study of the Chironomidae (Diptera) of Africa south of the Sahara. Part III. *Bulletin of the British Museum of Natural History (Entomology)* 5: 321–426. [Specimens, 3 males, 3 females collected Ruo, Nyasaland. No date.]

Rivellia woodi Frey (Brachycera, Platystomatidae)
Frey, R. 1932. On the African Platystomidae (Diptera). *Annals and Magazine of Natural History* 10(9): 242–264. [Specimens, 1 male collected Ruo, 4 December 1916; 1 male collected Limbe, Nyasaland, 24 December 1916.]

Chrysops woodi Neave (Brachycera, Tabanidae)
Neave, S. A. 1915. The Tabanidae of southern Nyasaland with notes on their life-histories. *Bulletin of Entomological Research* 5: 287–320. [Specimen, 1 female collected Chilanga, Northern Rhodesia, 4000ft. 1 December 1913.]

Chryosoma woodi Parent (Brachycera, Dolichopodidae)
Parent, O. 1935b Diptères Dolichopodides nouveaux. *Encycl. ent.* (B II) Dipt. 8: 59–96. [Specimen, Chilanga, Northern Rhodesia. 1913.]

APPENDIX 2

Name Changes

The place names in this book are those used when Wood lived in Africa. Since Independence, some names have changed as given below.

Previous Name	Present Name
Cholo	Thyolo
Florence Bay	Chitimba
Fort Hill	Chitipa
Fort Johnston	Mangochi
Fort Lister	Phalombe
Fort Manning	Mchinji
Lake Nyasa	Lake Malawi
Northern Rhodesia	Zambia
Nyasaland	Malawi
Salisbury	Harare
South-west Africa	Namibia
Southern Rhodesia	Zimbabwe
Tchiromo	Chiromo
Umtali	Mutare

BIBLIOGRAPHY

Bibliography (including all publications by R. C. Wood)

Andersen, K. (1914). A new *Nycteris* from N. W. Rhodesia. *Annals and Magazine of Natural History* 8(13): 563.

Anderson-Morshead, A.E.M. (1903). *The Building of the Chauncy Maples*. U.M.C.A., London. 124pp. (Reprinted in 1991 as *Lady of the Lake. The Story of Lake Malawi's M.V. Chauncy Maples* (with a history of the boat on the lake 1902–1990, by Vera Garland). Central Africana, Blantyre, Malawi. 161 pp.

Anon. (1906). Obituary: E. H. Howson. *The Harrovian* XIX(1): 2–4.

Anon. (1933). A.V. (Archer Vassall). An appreciation on retirement. *The Harrovian* 46: 66–68.

Anon, (1940). Obituary: Archer Vassall. *The Harrovian* LIV (11): 37.

Ansell, W.F.H., Benson, C.W. and Mitchell, B.L. (1962). Notes on some mammals from Nyasaland and adjacent areas. *Nyasaland Journal* 15(1): 38–54.

Ansell, W.F.H. and Dowsett, R.J. (1988). *Mammals of Malawi: An Annotated Checklist and Atlas*. Trendrine Press, Zennor, Cornwall. 170 pp + 53 pp. maps.

Arnold, C.W.B. (1952). Lake Nyasa's varying level. *Nyasaland Journal* 5(1): 7–17.

Baker, C. (1971). *Johnston's Administration: 1891–1897: A History of the British Central Africa Administration 1891–1897*. Department of Antiquities, Publication no. 9. Government Press, Zomba, Malawi. 134 pp.

Ballantyne, M.M.S. and Shepherd, R.H.W. (1968). *Forerunners of Modern Malawi: the early missionary adventures of Dr James Henderson 1895–98*. Lovedale Press, South Africa. 303 pp.

Belcher, C.F. (1930). *The Birds of Nyasaland*. Crosby Lockwood and Son, London. 356 pp. + map.

Benson, C.W. and Benson, F.M. (1977). *The Birds of Malawi*. Montfort Press, Limbe, Malawi. 263 pp.

Bertram, C.K., Borley, H.J.H. and Trewavas, E. (1942). *Report of the Fish and Fisheries of Lake Nyasa*. Crown Agents, London. 181 pp.

Best, A.A. and Raw, W.G. (1928). *Rowland Ward's Records of Big Game*. (15th edn., Africa). Rowland Ward Publications Ltd, Wood Green, London. 494 pp.

Betjeman, J. (1969). *Victorian and Edwardian London from Old Photographs*. Batsford Ltd, London.

Bezzi, M. (1917). New Ethiopian fruit-flies of the genus *Dacus*. *Bulletin of Entomological Research* 8: 63–71.

Bezzi, M. (1918). Notes on the Ethiopian fruit-flies of the family Trypaneidae other than *Dacus* (s.l.) (Dipt.)-II. *Bulletin of Entomological Research* 9: 13–46.

Bezzi, M. (1924). Further notes on the Ethiopian fruit-flies, with keys to all the known genera and species. *Bulletin of Entomological Research* 15: 73–118, 121–155.

Boeder, R.B. (1980). *Alfred Sharpe of Nyasaland: builder of empire*. Printed privately. Zomba. 152 pp.

Boulenger, G.A. (1896). Description of new fishes from the upper Shire River, British Central Africa, collected by Dr Percy Rendall, and presented to the British Museum by Sir Harry H. Johnston, K.C.B. *Proceedings of the Zoological Society of London* 1896: 915–920.

Boulenger, G.A. (1897). A list of reptiles and batrachians collected in northern Nyasaland by Mr. Alex. Whyte, F.R.S., and presented to the British Museum by Sir Harry Johnston, K.C.B.; with descriptions of new species. *Proceedings of the Zoological Society of London* 1897: 800–803.

Boulenger, G.A. (1902). Diagnoses of new cichlid fishes discovered by Mr J.E.S. Moore in Lake Nyasa. *Annals and Magazine of Natural History* 7(10): 69–71.

Boulenger, G.A. (1908). Diagnoses of new fishes discovered by Captain E.C. Rhoades in Lake Nyasa. *Annals and Magazine of Natural History* 8(9): 238–243.

Boulenger, G. A. (1909 - 1916). *Catalogue of the Fresh-water Fishes of Africa in the British Museum (Natural History)*. Trustees of the British Museum of Natural History, London. 4 vols. (Vol I [1909] 373 pp.; Vol II [1911] 529 pp.; Vol III [1915] 526 pp.; Vol IV [1916] 392 pp.).

Boy Scouts Association Canada. (1923). *Annual Report 1923*. Canadian General Council, Boy Scouts Association.

Boy Scouts Association Canada. (1924). *Annual Report 1924*. Canadian General Council, Boy Scouts of Canada.

Buchanan, J. (1885). *The Shirè Highlands as Colony and Mission*. William Blackwood and Sons, Edinburgh. 260 pp + map. (Reprint 1982; Blantyre Print and Publishing, Blantyre, Malawi).

Burke, T. (1940). *The Streets of London*. Batsford Ltd, London. 152 pp.

Butler, A.G. (1895). On a small collection of butterflies made by Consul Alfred Sharpe at Zomba, British Central Africa. *Proceedings of the Zoological Society of London* 1895: 720–721.

Buxton, P. A. (1953). *The Natural History of Tsetse Flies: an account of the Biology of the Genus* Glossina *(Diptera)*. H. K. Lewis, London. 816 pp + 27 plates.

Carr, A (1964). *Ulendo.* William Heinemann, London. 210 pp.

Carter, J. (1987). *Malawi - Wildlife, Parks and Reserves.* Macmillan, London. 176 pp.

Chadwick, O. (1959). *Mackenzie's Grave.* Hodder and Stoughton, London. 254 pp.

Charlton, P. (1993). Some notes on the Nyasaland Volunteer Reserve. *The Society of Malawi Journal* 46(2): 25–51.

Clendennen, G.W. (1994). Charles Livingstone and Malawi's first bird collection. *Society of Malawi Journal* 47(1): 39–60.

Cole-King, P.A. (1971). *Lake Malawi Steamers.* Department of Antiquities, Historical Guide Number 1. Government Printer, Zomba, Malawi. ca 40 pp. (not numbered).

Cole-King, P.A. (1972). Transport and communication in Malawi to 1891, with a summary to 1918. In: *The early history of Malawi.* Ed. B. Pachai. Longman, London. pp. 70–90.

Cole-King, P.A. (1973). *Blantyre: historical guide.* Christian Literature Association in Malawi. 25 pp.

Cole-King, P.A. (1982). *Cape Maclear.* Department of Antiquities, Publication No: 4. Government Printer, Zomba, Malawi. 67 pp.

Colville, A. (1911). *1000 Miles in a Machila.* Walter Scott Publishing Co. New York & Melbourne. 311 pp.

Coupland, R. (1928). *Kirk on the Zambesi.* Oxford University Press, Oxford. 286 pp.

Debenham, F. (1955). *Nyasaland: The Land of the Lake.* Her Majesty's Stationery Office, London. 239 pp.

de Guingand, F. (1953). *African Assignment.* Hodder and Stoughton, London. 291 pp.

de Meillon, B. (1930). On the Ethiopian Simulidae. *Bulletin of Entomological Research* 21(20): 185–200.

Dickson, C.G. C. (1974) (as C. G. C. D.) Obituary: K. M. Pennington (1897–1974*). Journal of the Entomological Society of Southern Africa* 37: 421–422.

Dickson, C.G.C. and Kroon, D.M. (eds) (1978). *Pennington's Butterflies of Southern Africa.* Ad. Donker (Pty) Ltd, Johannesburg. 670 pp +198 col pls + 1 map.

Dollman, J.G. and Borlase, J.B. (1928). *Rowland Ward's Records of Big Game.* (9th Edn.). Rowland Ward, Picadilly, London. 523 pp.

Dowsett-Lemaire, F. and Dowsett, R. 2006. *The Birds of Malawi.* Turaco Press & Aves, Liege, Belgium. 556 pp. + maps.

Drummond, H. (1889). *Tropical Africa.* Hodder and Stoughton, London. 228 pp.

Dudley, C. (1979). History of the decline of the larger mammals of the Lake Chilwa Basin. *Society of Malawi Journal* 32(2): 27–41.

Duff, H.C. (1906). *Nyasaland under the Foreign Office* (2nd ed.). George Bell & Sons, London. 422 pp.

Faulkner, H. (1868). *Elephant Haunts. A Sportsman's Narrative of the Search for Livingstone*. 324 pp. (Facsimile edition: Society of Malawi and Royal Geographical Society 1984).

Fort, G.S. (1932). *Alfred Beit*. Nicholson and Watson, London. 221 pp.

Foskett, R. (1965). *The Zambesi Journals and Letters of Dr John Kirk 1858–63*. 2 vols. Oliver and Boyd, Edinburgh. 636 pp. (Vol 1: xxi +1–317 pp.; Vol 2: 321–636 pp.).

Freeman, P. (1957). A study of the Chironomidae (Diptera) of Africa south of the Sahara, part III. *Bulletin of the British Museum of Natural History (Entomology)* 5: 321–426.

Freeman, P. and de Meillon, B. (1953). *Simuliidae of the Ethiopian Region*. Publication of the British Museum (Natural History) 194. 224 pp.

Fryer, G. and Iles, T.D. (1972). *The Cichlid Fishes of the Great Lakes of Africa: their biology and evolution*. Oliver and Boyd, Edinburgh. 641 pp.

Gelfand, M. (1964). *Lakeside Pioneers: Socio-medical study of Nyasaland (1875–1920)*. Basil Blackwell, Oxford. 340 pp.

Gifford, D. (1965). *A List of the Butterflies of Malawi*. The Society of Malawi, Blantyre, Malawi. 151 pp. + line drawings + 9 coloured plates.

Girling, B. (1993). *Images from Harrow: pictures from the past* (2nd impression). Dicker & Dunster Ltd., Mitcham, Surrey, U. K. 53 pp.

Glasgow, J. (1963). *Distribution and abundance of Tsetse*. Pergamon Press, Oxford. 241 pp.

Gunther, A. (1864). Report on a collection of reptiles and fishes made by Dr Kirk in the Zambesi and Nyasa regions. *Proceedings of the Zoological Society of London* 1864: 303–314.

Gunther, A. (1893). Second report on the reptiles, batrachians and fishes transmitted by Mr H.H. Johnston, C.B., from British Central Africa. *Proceedings of the Zoological Society of London* 1893: 616–628.

Hadlow, G.G.S. (1939). A short history of tea planting in Nyasaland. *Quarterly Journal of the Nyasaland Tea Association* 3: 15–19; 4(1): 7–11; 4(2): 3–7.

Hadlow, G.G.S. (1960). The history of tea in Nyasaland. *Nyasaland Journal* 13(1): 21–31.

Hancock, D. (1984). The princeps nireus group of swallowtails (Lepidoptera: Papilionidae). Systematics, phylogeny and biogeography. *Arnoldia Zimbabwe* 9: 181–215.

Hanna, A.J. (1956). *The Beginnings of Nyasaland and North-eastern Rhodesia 1859–95*. Oxford University Press, Oxford. 281 pp.

Hanney, P. (1965). The Muridae of Malawi. *Journal of Zoology, London* 146: 577–633.

Happold, M. and Happold D.C.D. (1997). New records of bats (Chiroptera: Mammalia) from Malawi, east-central Africa, with an assessment of their status and conservation. *Journal of Natural History* 31: 805–836.

Happold, M. and Happold, D.C.D. (1998a). Chiromo and Thyolo revisited:

comments on the conservation of small mammals in Malawi. *Nyala* 20:1–10.

Happold, M. and Happold, D.C.D. (1998b). New distribution records for bats and other mammals of Malawi. *Nyala* 20: 17–24.

Hayes, G.D. (1967). Nyala and the Lengwe Game Reserve. *Society of Malawi Journal* 20(2): 26–29.

Hayes, G.D. (1972). Wild Life Conservation in Malawi. *Society of Malawi Journal* 25: 22–31.

Hayes, G.D. (1978). *A Guide to Malawi's National Parks and Game Reserves*. Montford Press, Blantyre. 166 pp.

Hetherwick, A. (1931). *The Romance of Blantyre: How Livingstone's dream came true*. James Clarke & Co., London. 260 pp.

Hetherwick, A. (1962). *The Building of Blantyre Church, Nyasaland, 1888–1891* (3rd ed.). Blantyre Synod Bookshop, Nyasaland. 23 pp.

Hinton, M.A.C. (1921). Some new African mammals. *Annals and Magazine of Natural History* 9(7): 368–373.

Huxley, E. (1974). *Livingstone and his African Journeys*. Weidenfeld and Nicolson, London and Saturday Review Press, New York. 224 pp.

Ireland, S.J. (1983). *The School that We Love on the Hill*. Burns & Harris Limited, Dundee. 137 pp.

Jack, J.W. (1900). *Daybreak in Livingstonia*. Young People's Missionary Movement, New York). 371 pp. (reprint 1969: Negro University Press, New York.)

Jackson, P.B.N. (1961). Checklist of the fishes of Nyasaland. *Occasional Papers of the National Museums of Southern Rhodesia* 3(25B): 535–621.

Jeal, T. (1973). *Livingstone*. Heinemann, London. 427 pp.

Johnson, W.P. (1923). *My African Reminiscences 1875–1895*. Universities Mission to Central Africa. Westminster, London. 236 pp.

Johnston, F. and Garland, V. (1993). *Malawi. Lake of Stars*. Central Africana Ltd., Blantyre, Malawi. 160 pp

Johnston, H.H. (1897). *British Central Africa*. London: Methuen & Co, London. 544 pp.

Johnston, H.H. (1923). *The Story of My Life*. Chatto and Windus, London. 536 pp.

Kershaw, P.S. (1922). On a collection of mammals from Chiromo and Cholo, Ruo, Nyasaland, made by Mr Rodney C. Wood, with field-notes by the collector. *Annals and Magazine of Natural History* 9(10): 177–192.

King, M.E. (n.d. [*c*.1993]). *The Story of Medicine and Disease in Malawi*. Montfort Press, Blantyre, Malawi. 183 pp.

Kirk, J. (1864). List of Mammalia met with in Zambesia. *Proceedings of the Zoological Society of London* 1864: 649–660.

Konings, A. (1990). *Ad Konings's Book of Cichlids and all the Other Fishes of Lake Malawi*. TFH Publications Inc, Neptune City, USA. 495 pp.

Laws, R. (1934). *Reminiscences of Livingstonia*. Oliver and Boyd, Edinburgh. 272 pp.

Lewis, D., Reinthal, P. and Trendall, J. (1986). *A Guide of the Fishes of Lake*

Malawi National Park. World Wildlife Fund, Gland, Switzerland. 71 pp.

Lewis, D.J. (1960). Observations on the *Simulium neavei* complex at Amani in Tanganyika. *Bulletin of Entomological Research* 51: 95–113.

Lewis, D.J. and Hanney, P.W. (1965). On the *Simulium neavei* complex. *Proceedings of the Royal Entomological Society of London (B)* 34: 12–16.

Lionnet, G. (1972). *The Seychelles*. Wren Publishing, Melbourne. 200 pp.

Livingstone, W.P. (1921). *Laws of Livingstonia*. Hodder and Stoughton, London. 385 pp.

Lorenz, F. Jr. and Hubert, A. (1993). *A Guide to Worldwide Cowries*. Verlag Christa Hemmen, Wiesbaden. 571 pp.

Loveridge, A. (1949). *Many Years I've Squandered*. Scientific Book Club, London. 243 pp.

Loveridge, A. (1951). *Tomorrow's a Holiday*. Robert Hale, London. 303 pp.

Loveridge, A. (1953). *I Drank the Zambesi*. Harper and Brothers, New York. 206 pp.

Lowe-McConnell, R. (2003). Recent research in the African Great Lakes: Fisheries, biodiversity and cichlid evolution. *Freshwater Forum* 20: 1–64.

Lowe-McConnell, R. (2006). *The Tilapia Trail – the life of a fish biologist*. MPM Publishing. Ascot, England. 296 pp + illus.

Martelli, G. (1970). *Livingstone's River: A history of the Zambesi expedition 1858–1864*. Chatto and Windus, London. 260 pp.

Maugham, R.C.F. (1929). *Africa as I Have Known it*. John Murray, London. 372 pp.

Maydon, H.C. (1932). *Big Game Shooting in Africa*. Lonsdale Library no. 14. Seeley, Service and Co., London. 445 pp.

McDonald, L.J. (1961). Bishop Mackenzie's grave. *Society of Malawi Journal* 14(1): 60–67.

McIntosh, K. and Norton, L. (1987). *Echoes of Enterprise*. Enterprise Farmers Association, Highlands, Zimbabwe. 232 pp.

Medland, B and Zalinge, N. P. (n. d.). *Checklist of the birds seen in the Lake Malawi National Park and its immediate surroundings*. Wildlife Society of Malawi. Lilongwe, Malawi. 11 pp.

Mell, A.H. (1960). *History of the Old Residency and Government House*. Government Printer, Zomba, Malawi. 10 pp + photos.

Mitchell, B.L. (1946). A vocabulary giving the English, scientific and African names for some of the animals in Nyasaland. *Nyasaland Agricultural Quarterly Journal* 6: 25–47.

Mitchell, P.C. (1929). *Centenary History of the Zoological Society of London*. London: Zoological Society of London.

Moir, F.L.M. (1923). *After Livingstone: An African trade romance*. Hodder and Stoughton, London. 200 pp.

Morris, B. (1995). Wildlife depredation in Malawi: the historical dimension. *Nyala* 18: 17–24.

Morris, B. (1996). Conservation mania in colonial Malawi: another view. *Nyala* 19: 17–26.

Morris, B. (1997). G.D. Hayes and the Nyasaland Fauna Preservation Society. *Society of Malawi Journal* 50(1): 1–12.

Morris, M. (1952). *A Brief History of Nyasaland.* Longmans, Green & Co., London. 65 pp.

Moyse-Bartlett, H. (1956). *The King's African Rifles. A Study in Military History of East and Central Africa, 1890–1945.* Gale and Polden, Aldershot, U.K. 727 pp.

Muldoon, G. (1955). *Leopards in the Night.* Hart-Davies, London. 234 pp.

Murray, S.S. (1932). *A Handbook of Nyasaland.* : Crown Agents for the Colonies, London. 436 + xxxvii pp.

Neave, S.A. (1915). The Tabanidae of southern Nyasaland with notes on their life-histories. *Bulletin of Entomological Research* 5: 287–320.

Oliver, R. (1964). *Sir Harry Johnston and the Scramble for Africa.* Chatto and Windus, London. 368 pp.

Parent, O. (1935). Diptères Dolichopodides nouveaux. In *Encyclopédie Entomologique. Série B. Diptera: Parasitologie, Biologie, Systématique. Tome VIII.* Eds. M. Goetghebuer, M.A. Jourdan, O. Parent, F. Quivreux & E. Seguy. Paul Le chevalier, Paris. pp. 59–96.

Perham, M. (1956). *Lugard: The years of adventure 1858–1898.* Collins, London. 750 pp.

Pike, J.G. and Rimington, G.T. (1965). *Malawi - a geographical study.* Oxford University Press, London. 229 pp.

Pinhey, E. (1966). Checklist of the dragonflies (Odonata) from Malawi with description of a new *Teinobasis* Kirby. *Arnoldia* 33(2): 1–24.

Pinhey, E.C.G. (1962). Rodney Carrington Wood. (obituary). *Occasional Papers of the National Museums of Southern Rhodesia* 26: 912–913.

Potous, P.L. (1956). *No Tears for the Crocodile.* Hutchinson & Co, London. 188 pp.

Ransford, O. (1966). *Livingstone's Lake: The drama of Nyasa.* John Murray, London. 313 pp.

Raybould, J.N. (1969a). Studies on the immature stages of the *Simulium neavei* Roubaud complex and their associated crabs in the Eastern Usumbara Mountains in Tanzania. *Annals of Tropical Medicine and Parasitology* 63: 269–287.

Raybould, J.N. (1969b). A study of the anthropophilic female Simuliidae (Diptera) at Amani, Tanzania: The feeding behaviour of *Simulium woodi* and the transmission of onchocerciasis. *Annals of Tropical Medicine and Parasitology* 63: 76–88.

Rees, J. D. (c 1910). *Nyasaland and the Shiré Highland Railway.* British Central Africa Company Limited, London. 54 pp (1986 Reprint, with introduction and history 1910–1986: Montfort Press, Limbe, Malawi.)

Regan, C.T. (1921). The cichlid fishes of Lake Nyassa. *Proceedings of the Zoological Society of London* 21: 675–727.

Reynolds, D. (1997). *Steam and Quinine on Africa's Great Lakes.* Bygone Ships, Trains & Planes. Pretoria, South Africa. 217 pp.

Reynolds, E. E. (1960). *The Scout Movement.* Oxford University Press, London. 234 pp,

Richer, C. (1951). *Ti-coyo and His Shark.* Alfred A. Knopf, New York. 235 pp.

Richer, C. (1954). *Son of Ti-coyo.* Alfred A. Knopf, New York. 245 pp.

Riley, N.D. (1943). A new African lycaenid butterfly. *Entomologist* 76: 225–226.

Sclater, P.L. (1892). [On a small collection of mammals brought by Mr. A. Sharpe from Nyasaland]. *Proceedings of the Zoological Society of London* 1892: 97–98.

Sclater, P.L. (1893). [On skins of mammals from Nyasaland]. *Proceedings of the Zoological Society of London* 1893: 506–507.

Sclater, P.L. (1896). [On the Nyasaland Gnu, *Connochaetes taurinus johnstoni*]. *Proceedings of the Zoological Society of London* 1896: 616–618.

Shepperson, G. and Price, T. (1958). *Independent African: John Chilembwe and the origins, setting and significance of the Nyasaland native rising of 1915.* Edinburgh University Press, Edinburgh. 564 pp.

Skerrett, J. and Skerrett, A. (Eds) (1998). *Seychelles* (2nd ed., revised). Insight Pocket Guides, APA Publications. 94 pp.

Smithers, R.H.N. (1984). Recollections of some great Naturalists. *Transvaal Museum Bulletin* 20: 5–15.

Stigand, C.H. and Lyell, D.D. (1906). *Central African Game and its Spoor.* Horace Cox, London. 315 pp.

Stuart-Mogg, D. (1994). *A Guide to Malawi.* Central Africana Ltd, Blantyre, Malawi. 106 pp. + illus.

Sweeney, R.C.H. (1959). *A Checklist of the Mammals of Nyasaland.* The Nyasaland Society, Blantyre, Malawi. 71 pp.

Tattersall, D. (1982). *The Land of the Lake. A Guide to Malawi.* Blantyre Periodicals, Blantyre, Malawi. 176 pp.

Tea Association of Malawi (1986). *Tea Vignettes - 50th Anniversary 1936–1986.* Tea Association of Malawi, Blantyre, Malawi. 27 pp. + 24 drawings by D. Brian Roy.

Thomas, O. (1892). On mammals from Nyasaland. *Proceedings of the Zoological Society of London* 1892: 546–554.

Thomas, O. (1893). On a second collection of mammals sent by Mr. H.H. Johnston, C.B., from Nyasaland. *Proceedings of the Zoological Society of London* 1893: 500–504.

Thomas, O. (1894). On the mammals of Nyasaland: third contribution. *Proceedings of the Zoological Society of London* 1894: 136–146.

Thomas, O. (1896). On the mammals of Nyasaland: fourth notice. *Proceedings of the Zoological Society of London* 1896: 788–798.

Thomas, O. (1897a). [Exhibition of specimens and descriptions of new species of mammals from northern Nyasaland, with a note on the genus *Petrodromus*].

Proceedings of the Zoological Society of London 1897: 430–436.

Thomas, O. (1897b). On the mammals obtained by Mr. A. Whyte in Nyasaland, and presented to the British Museum by Sir H.H. Johnston, KCB; being a fifth contribution to the mammal-fauna of Nyasaland. *Proceedings of the Zoological Society of London* 1897: 925-939.

Thomas, O. (1898). On a small collection of mammals obtained by Mr Alfred Sharpe, C.B. in Nyasaland. *Proceedings of the Zoological Society of London* 1898: 391–394.

Thomas, O. (1902). On some new mammals from northern Nyasaland. *Proceedings of the Zoological Society of London* 1902: 118–121.

Thomas, O. (1917). A new species of *Scotoecus*. *Annals and Magazine of Natural History* 8(19): 280–281.

Trewavas, E. (1935). A synopsis of the cichlid fishes of Lake Nyasa. *Annals and Magazine of Natural History* 10(16): 65–118.

van Riet Lowe, C. (1936). The Smithfield 'N' Culture. *Transactions of the Royal Society of South Africa* 23(4): 367–372.

van Riet Lowe, C. (1941). Bored stones in Nyasaland. *South African Journal of Science* 37: 320–326.

van Someren, V.G.L. (1964). Revisional notes on African Charaxes (Lepidoptera: Nymphalidae), part II. *Bulletin of the British Museum (Natural History) Entomology* 15(7): 181–235 + 23 plates, 5 maps.

Vine, P. (1989). *Seychelles*. Immel Publishing, London. 209 pp.

von Lettow-Vorbeck, P. (1922). *My Reminiscences of East Africa*. Hurst and Blackett. 336 pp.

Westrop, A. (1962). Rodney Carrington Wood. [obituary]. *Nyasaland Journal* 15(2): 38–39.

Westrop, A. (n.d.). *Green Gold*. Printed privately. 491 pp.

Whitehouse, H.E. (1952). Level of Lake Nyasa. *Society of Malawi Journal* 5(1): 18–21.

Wilson, G.H. (1936). *The History of the Universities' Mission to Central Africa*. Universities' Mission to Central Africa, London. 278 pp.

Winspear, F. (1960). *Some Reminiscences of Nyasaland*. Universities Mission Press, Likoma, Nyasaland. 45 pp.

Witherby, H. F. (1943). Hubert Lynes: a biographical sketch. *Ibis* 85: 198–215.

Wood, R.C. (1926a). 'The exception' - and a pair of wagtails. *Nyasaland Times* 30 March.

Wood, R.C. (1926b). Gun-boys. *Nyasaland Times* 28 May.

Wood, R.C. (1927–1960). Diaries, 9 vols. Cambridge University Library (catalogued as: Diaries of Rodney Wood, GBR/0115/RCMS 128).

Wood, R.C. (1927a). The apparent crepuscular dances of two species of African butterflies [communicated by Dr S.A. Neave]. *Proceedings of the Royal Entomological Society of London* 2: 42–43.

Wood, R.C. (1927b). Some hints on the conservation of our bird life.

Unpublished [headed 'For "Nyasaland Times"'], 17 February.

Wood, R.C. (1929). Observations by Rodney C. Wood on butterflies and their chief vertebrate pests in Nyasaland. [communicated by Professor Poulton]. *Proceedings of the Royal Entomological Society of London* 4: 40–41.

Wood, R.C. (1932a). Nyasaland 1: General. In *Big Game shooting in Africa*. Ed. H.C. Maydon. Seeley, London (Lonsdale Library no. XIV). pp. 315–320.

Wood, R.C. (1932b). Nyasaland 2: Nyala hunting. In *Big Game Shooting in Africa*. Ed. H.C. Maydon. Seeley, London. (Lonsdale Library no. XIV). pp. 321–327

Wood, R. C. (1933). Birds - allies, not enemies, of man. *Natal Mercury* 19 June: 22–23.

Wood, R.C. (1935a). Another note on *Sarothrura elegans*. *Ostrich* 6: 105–106.

Wood, R.C. (1935b). Natural History Society. *Michaelhouse Chronicle* 1935: 18–20.

Wood, R.C. (1936). Natural History Society. *Michaelhouse Chronicle* 1936: 20–21.

Wood, R.C. (1937a). A note on the behaviour of the Nyasaland Coucal (*Centropus burchellii fasciipygialis*). *Ostrich* 8: 40–41.

Wood, R.C. (1937b). The remarkable 'false face' of the Pearl-spotted Owlet. *Ostrich* 8: 114–115.

Wood, R.C. (1940). Drumming of the Black-throated Honey-guide (*Indicator indicator*). *Ostrich* 11: 50–51.

Wood, R.C. (1941). A note on the Crowned Crane (*Balearica regulorum*). *Ostrich* 12: 26–27.

Wood, R.C. (1942a). Birds caught in spiders' webs in Nyasaland. *Ostrich* 13: 47–49.

Wood, R.C. (1942b). A cuckoo problem. *Ostrich* 13: 172–173.

Wood, R.C. (1942c). Nesting habits of the Knob-billed Goose (*Sarkidiornis melanonotus*). *Ostrich* 13: 173–176.

Wood, R.C. (1943). The food and distribution of the Vulturine Eagle. *Ostrich* 14: 107–108.

Wood, R.C. (1946). Crowing Crested Cobras. *Nyasaland Times*, 3 June.

Wood, R.C. (1949a). 'Bushman occupation' at Monkey Bay. *Nyasaland Journal* 2(1): 57.

Wood, R.C. (1949b). Outstanding trees. *Nyasaland Journal* 2(1): 15.

Wood, R.C. (1949c). A rare butterfly - Gynandromorph *Cupido nyasae* (Lycaenidae). *Nyasaland Journal* 2(1): 5.

Wood, R.C. (1949d). Small mammals of southern Nyasaland. *Nyasaland Journal* 2(1): 38–41.

Wood, R.C. (1951). Stone Age cultures in Nyasaland. *Nyasaland Journal* 4(1): 65–66.

Wood, R.C. (1958a). An anecdote of Livingstone's first landing in Nyasaland. *Nyasaland Journal* 11(2): 39–40.

Wood, R.C. (1958b). 'Elephant' island on Lake Nyasa. *Nyasaland Journal* 11(1): 23–24.

Wood, R.C. (1958c). On the distribution of the butterfly *Cymothoe theobene* in Nyasaland. *Nyasaland Journal* 11(2): 28.

Wood, R.C. (n.d.). Conservation of bird life. (Typed manuscript not published.)

Worthington, E.B. (1933). The fishes of Lake Nyasa (other than Cichlidae). *Proceedings of the Zoological Society of London* 1933: 285–316.

Zoological Society of London (1895). Report of the Council for the year 1894. Medallists, p. 5.

Zoological Society of London (1898). Report of the Council for the year 1898. Medallists, p. 5.

INDEX